Mass Psychogenic Illness: A Social Psychological Analysis

ENVIRONMENT AND HEALTH

A series of books edited by
Andrew Baum and **Jerome E. Singer**

BAUM and EPSTEIN • *Human Response to Crowding, 1978*

COLLIGAN, PENNEBAKER, and MURPHY • *Mass Psychogenic Illness: A Social Psychological Analysis, 1982*

MASS PSYCHOGENIC ILLNESS:
A Social Psychological Analysis

Edited by
Michael J. Colligan
Miami University
Oxford, Ohio

James W. Pennebaker
University of Virginia
Charlottesville, Virginia

Lawrence R. Murphy
Xavier University
Cincinnati, Ohio

Routledge
Taylor & Francis Group
NEW YORK AND LONDON

First Published by
Lawrence Erlbaum Associates, Inc., Publishers
365 Broadway
Hillsdale, New Jersey 07642

Transferred to Digital Printing 2009 by Routledge
270 Madison Avenue, New York, NY 10016
27 Church Road, Hove, East Sussex BN3 2FA

Library of Congress Cataloging in Publication Data

Main entry under title:

Mass psychogenic illness.

(Health psychology)
Includes bibliographies and indexes.
1. Hysteria—Psychosomatic aspects. 2. Hysteria (Social psychology) 3. Industrial psychiatry.
I. Colligan, Michael, II. Pennebaker, James W.
III. Murphy, Lawrence R. IV. Series. [DNLM: 1. Psychophysiologic disorders. 2. Mass behavior. WM 90 M414]
RC532.M37 616.08 82-5052
ISBN 0-89859-160-0 AACR2

Publisher's Note
The publisher has gone to great lengths to ensure the quality of this reprint but points out that some imperfections in the original may be apparent.

Contents

Preface

The spontaneous outbreak of illness among a group of workers is a dramatic and distressing event. Of primary concern, of course, is the immediate treatment and recovery of the affected workers and the detection and elimination of the illness source. To the extent that the cause can be readily identified and understood, anxiety surrounding the event diminishes. Once the affected individuals have a rational explanation for what happened to them, they can begin to realistically evaluate their symptoms, not only in terms of their immediate significance, but also in terms of their longrange implications.

Unfortunately not all incidents of mass illness in work settings are so easily resolved. In some cases a physical or chemical cause for the illness may not be obvious. Recurrent episodes may occur long after the initiating cause has been removed or the expressed symptoms may be too nonspecific and transient to permit diagnosis. Under these conditions the illness remains a mystery and the workplace becomes a breeding ground for anxiety, rumor, confusion, and fear. Economic pressures to resume normal operations are counteracted by genuine concerns that the environment is still pathogenic. Having exhausted all possible physical explanations for the illness episode, investigators may turn to a psychological interpretation of the outbreak, explaining it in terms of mass hysteria or mass anxiety. This conclusion is always a tenuous one, being based primarily on the absence of physical evidence rather than the presence of a clearly defined set of precipitating psychosocial conditions. This ambivalence is understandable given the long and controversial history of the mind-body issue in general, and the concept of "hysteria" in particular. Despite these semantic and theoretical pitfalls, the fact remains that the workplace is as much a psychosocial environment as it is a physical one. The effects of psychogenic stressors may be as threatening and debilitating to a workforce as excessive noise levels or inadequate ventilation.

There is an urgent need to reexamine the old issue of mass hysteria in terms of contemporary theory and research findings. By what mechanisms do psychological properties and processes become translated into physical symptoms? What conditions promote or limit the contagion of psychogenic symptoms throughout a workforce? Are some individuals or work settings more susceptible to such illness outbreaks than others, and if so what are these critical precipitating conditions? Is it possible that predisposing psychological factors may, in and by themselves, trigger a contagion of psychogenic symptoms? To what extent do these same psychological stressors lower worker resistance to sub-standard chemical exposures, making them hypersensitive to mild irritants in the work environment? What steps can be taken to contain the spread of panic and anxiety once a mass illness outbreak has occurred? Complexity is not a viable excuse for avoiding these issues, and while we might not be in a position to apply any quick and easy solutions to these problems we can at least attempt to develop a conceptual framework for understanding them.

The present book brings together scientists from several disciplines in an attempt to answer some of these questions. Because mass psychogenic illness has rarely been systematically examined, it is necessary to explore outbreaks from a variety of perspectives, including historical, cultural, social, psychological, and even medical. Obviously, no single viewpoint can adequately explain this phenomenon. Nevertheless, it is our hope that researchers, physicians, management and workers can begin to understand which variables in the natural environment are related to the occurrence and amelioration of mass psychogenic illness.

This volume is divided into three sections. The first part is intended to give the reader a flavor of the many forms that mass psychogenic illness has taken. As the reader will learn, contagion of psychogenic illness has occurred throughout history and across cultures. The symptoms and behaviors have differed, but the psychological processes appear to be similar. The first section of the book will also point to some dramatic case studies that have been extensively researched within the United States in recent years in factory settings. These studies will point to the temporal development of an outbreak and suggest some of the complexities involved in studying mass psychogenic illness.

The second section of the book deals with some of the methodological problems associated with mass psychogenic illness. The researcher in this field faces several difficulties, such as whether to look at the phenomenon on an individual or organizational level. Inherent biases and linguistic pitfalls associated with adopting either a management or worker perspective can easily compromise the researcher's efforts no matter how well-intentioned they may be. Within this section, the reader will learn where many of the potential problems lie and how to avoid them.

The final section of the book addresses many of the theoretical underpinnings of mass psychogenic illness. Because the phenomenon essentially occurs on a group level among individuals, explanations must take into account psychological stressors, physiological changes, perceptions of body and environment, friendship networks, and a host of other critical dimensions. Within this section, then, each author analyses the phenomenon with a unique perspective. Consequently, each approach supplements other approaches. The converging viewpoints will aid individuals who are interested in either the theoretical or applied problems to better understand the many facets of mass psychogenic illness. Although our theories concerning mass psychogenic illness are nascent, they provide a beginning framework by which we can learn to prevent and control outbreaks of psychogenic illness.

We must emphasize that although this book deals specifically with mass psychogenic illness, it has direct implications for a variety of social and psychological phenomena. Readers with interests in community psychology, psychosomatics, group behavior, or even business and management will find most of the chapters useful and relevant. Finally, our focus on the psychodynamics of mass illness episodes does not imply a disregard for the physical and chemical hazards that potentially exist in the workplace. Rather, our aim is to complement many of the traditional evaluative processes in the investigation of a mass illness outbreak.

The current volume grew out of a symposium sponsored by the National Institute for Occupational Safety and Health (NIOSH) dealing with the diagnosis and amelioration of mass psychogenic illness. In an effort to supplement many of the viewpoints advanced at the conference, additional chapters were prepared by several of the authors. We wish to thank Elliott Harris for his continual help on this project. Finally, we are indebted to NIOSH for their financial support of the original symposium.

Michael J. Colligan
James W. Pennebaker

Mass Psychogenic Illness:
A Social Psychological Analysis

1 OVERVIEW OF MASS PSYCHOGENIC ILLNESS: HYSTERIA REVISITED?

The notion that psychosocial factors may play an important role in the etiology of certain episodes of mass illness is hardly new. Reports of such incidents are threaded throughout the history of civilization, providing a reflection of the sociocultural and philosophical currents of the times. It was Hippocrates, writing in about 400 B.C., who first introduced the term ''hysteria'', derived from the Greek word ''hysterikos'' meaning ''womb,'' to explain the appearance of a wide range of symptoms (e.g., convulsions, twitching, muscle spasms, headache, abdominal cramps) in unmarried Greek women. It was the contention of Hippocrates that sexual abstinance was an unnatural state, agitating the womb and causing it to wander throughout the body in search of satisfaction. The meanderings of the womb produced discomfort in the affected regions, and resulted in an overall excited state in the victim (Veith, 1965). Presumably this affliction could spread to other women in the vicinity of the victim, creating an epidemic reaction or ''mass hysteria.''

With the advent of the dark ages and the pervasive concern with religiosity and righteousness, outbreaks of mass psychogenic episodes became more frequent. During this time, it was commonly believed that such afflictions resulted from the demonic possession of unworthy souls. Perhaps the best known example of such episodes is the St. Vitus Dance epidemics that swept through Europe in the fifteenth and sixteenth centuries. Martin (1923) attributes the origin of this phenomenon to a popular mid-evil fable involving an event rumored to have occurred in a church-side graveyard in the German countryside on Christmas night in 1021. A group of pagens, consisting of both men and women, had met to celebrate Christmas in a sacriligous night of song and dance when their revelry disturbed the services of the parish priest in the nearby church. His appeals to them to stop proved unsuccessful. In his anger, the priest placed a curse on them that they should dance without rest for a full year. The rumor goes that the wretched souls had danced themselves into the ground up to their knees before two bishops happened by, and, sympathetic to their pleas, removed the curse. Nevertheless, the ordeal had been so traumatic as to leave them trembling and spasmatic for the rest of their lives. In appearance, it was said that their uncoordinated gait resembled that of the beloved St. Vitus who had apparently suffered from neuropathy. At any rate, the curse of St. Vitus became a popular malediction which was levied against enemy and sinner during the dark ages. It was widely believed that the guilty would be physically punished for their shortcomings. Not surprisingly, this prophecy was fulfilled and mass outbreaks of uncontrolled dancing, hopping, and jumping were periodically reported throughout the countryside.

In 13th century Italy, a particular form of St. Vitus Dance involving spasms, uncontrolled vocalizations, and dancing till exhaustion was thought to be caused by the bite of a spider, the tarantula. It was not uncommon for a victim of "tarantism" to rush naked from his or her home to dance and shout in the streets. Often the victim would be joined by neighbors until, by some counts, as many as 500 individuals were similarly affected (Martin, 1923). Once bitten, the victims believed their systems to be permanently poisoned, so that they were subject to relapses for the rest of their lives.

The dynamics and triggering factors of the St. Vitus Dance episodes have recurred throughout history and have been implicated in mass outbreaks of illness, halucinations, and a variety of behaviors described in the first chapter of this section. The "spells" and trance states reported by the victims of recent mass illness outbreaks in Singapore described by Dr. Phoon in Chapter 2 are reminiscent of the bizarre symptoms characterizing the St. Vitus Dance epidemics. An objective review of past cases can do much to enhance our understanding of the role of psychosocial factors in the etiology of illness. Incidents such as these have been recorded for centuries among a variety of cultures within factories, schools, convents, and other social organizations. Not only have these outbreaks spanned time and culture, but the symptoms and the behaviors associated with them have differed tremendously. Where one culture may typically have mass illness episodes characterized by symptoms of nausea, headaches and dizziness, other cultures may have outbreaks wherein the affected individuals display common halucinations or violent behaviors. As the following chapters will suggest, the differences in the symptoms can partially be explained by the prevailing cultural, religious, and scientific beliefs held by the society. Despite these large differences, a number of remarkable similarities appear to underly their occurrence. For example, most incidences occur among individuals in stressful environments who have friends who have also been affected. Although the triggering event may differ from outbreak to outbreak, the general pattern of the contagion of symptoms is similar. Clearly, the similarities among the outbreaks are as intriguing as the differences between them.

Finally, although our appreciation and understanding of mass psychogenic illness can be enhanced by an historical perspective, we should not feel constrained by the assumptions of our predecessors. The term "hysteria" has become so laden with surplus meaning and controversy that one frequently confuses this label with the phenomenon itself, and in rejecting the former, one loses sight of the latter. One of the primary aims of this volume is to reexamine outbreaks of "mass hysteria" in terms of current social psychological theory and knowledge.

The first section of the book is intended to acquaint the reader with the basic phenomenon of mass psychogenic illness. Each chapter offers a unique view of documented cases. The first chapter, written by Dr. Kerckhoff, gives an indepth account of one particular case that occurred in 1962 in a textile mill. In his account of the classic June Bug outbreak, the evolving psychological and sociological processes within a typical case are carefully documented. Dr. Phoon's chapter presents several case studies that have occurred within various factories in Singapore. His account offers a startling example of cultural and subcultural influences on incidents of mass illness episodes. The final chapter, by Drs. Colligan and Murphy, summarizes several outbreaks of mass illness within factories in the United States. The authors point to several common situational and personality factors implicated in the contagion of illness.

It must be noted that the reader interested in additional case studies will find numerous examples in several of the later chapters related to the theoretical issues surrounding mass psychogenic illness (Section 3 of the book). For example, Dr. Sirois lists a large number of hysteria outbreaks that have occurred over the centuries as they relate to his psychodynamic orientation (chapter 13). In addition, Dr. Stahl—in chapter 11—chronicles an interesting illness outbreak that occurred in a computer center. Shorter references to other such cases are discussed in virtually all of the remaining chapters.

REFERENCES

Martin A. History of dancing mania. *American Journal of Clinical Medicine*, 1923, *30*, 265–271.

Veith, I. *Hysteria: The history of a disease*. Chicago: University of Chicago Press, 1965.

1 Analyzing a Case of Mass Psychogenic Illness

Alan C. Kerckhoff
Duke University

What follows is a brief account of a study of mass psychogenic illness which was originally published in book form under the title *The June Bug*. Most of what is presented here is in the form of excerpts from that volume, chosen so as to encompass many of the more important elements in the event itself and in our analysis of it. Further details may be found in the book.[1]

Given the nature of such cases, we knew nothing about the event until its report by the mass media. It was thus practically over before we even made contact with the company where it occurred. As so often happens, therefore, we were necessarily dealing with the problem of trying to understand the case after it was over, based on what few records existed and on interview data that could be collected later on. It is always possible to interpret the results, therefore, as of questionable validity. For instance, though we found that persons affected in this case were different in significant ways from those who were not affected, one might argue that the differences were *due to* different kinds of participation in the event rather than *explanations of* that participation. Although we are persuaded that our results form a reasonable explanation of participation, significant methodological issues remain.

THE "BUG" AND THE EPIDEMIC

Word first reached the public on the 6 o'clock news. The report was brief, and an air of mystery was already evident:

---

[1] *The June Bug: A Study of Hysterical Contagion,* by Alan C. Kerckhoff and Kurt W. Back, 1968, New York: Appleton-Century-Crofts. I am grateful to Kurt Back and to Prentice-Hall, Inc., current holders of the copyright, for permission to publish these portions of the book.

> Officials of Montana Mills shut down their Strongsville plant this afternoon because of a mysterious sickness.
>
> According to a report just in from Strongsville General Hospital, at least ten women and one man were admitted for treatment. Reports describe symptoms as severe nausea and a breaking out over the body.
>
> Indications are that some kind of insect was in a shipment of cloth that arrived from England at the plant today. And at the moment the bug is blamed for the outbreak of sickness.

Later that night, on the 11 o'clock news, further details were supplied. Some of them varied from the original report, but the melodramatic tone remained:

> During the past three weeks a number of the 200 employees have been stricken with a mysterious illness, apparently caused by an insect bite. Today about ten women and one man were stricken. Several were admitted to the hospital for treatment and observation. Company officials say they are fumigating the building.
>
> The plant is scheduled to reopen tomorrow morning at six o'clock.
>
> This station learned tonight from a company employee that the small insect attacks the skin, the bites leaving a wound similar to a gnat bite. In about twenty minutes the victim is struck with severe nausea. The company doctor informed us tonight that an entomologist is studying the problem. A report is expected later this week.

It was a Wednesday night in June, 1962. . . . Before it ended, the story became considerably more complex and the cast of players grew markedly. Some feeling for the event and the reactions to it can be derived from a brief sampling of the reports of the various news media over the next few days:

Thursday— . . . Hiram L. Lamont, plant personnel director, said a number of women reported to work today, when the plant resumed operations, in a highly nervous state. At least six were treated by the company physician and sent home.

Dr. C. H. Foreman, Strongsville County Health Officer, reports tonight, however, that there is nothing present in the community to get excited about. . . . The predominating symptom according to physicians and company officials is anxiety. . . . Dr. Foreman says the doctors have ruled out a virus—since none have fever—and food poisoning. All are in good condition.

Friday—Two experts from the U.S. Public Health Service Communicable Disease Center arrived today in Strongsville to assist local health officials trying to determine the cause of the sudden outbreak of sickness which has hit employees of Montana Mills. . . . Also on the premises were several plant officials, representatives of the plant's insurance company, two State College entomologists, representatives of the Strongsville Exterminating Company, and an engineer from the State Board of Health. . . .

Nine persons remained hospitalized last night and two more have been treated in the hospital emergency room. There were unconfirmed reports this morning that at least four more persons have been hospitalized. About fifty persons have been affected. . . .

Dr. Foreman said that the cause of the illness is still unknown. The two Public Health Service doctors are returning to the Communicable Disease Center tonight with several specimens found today. The specimens are identified as small insects about the size of a mite.

Baffled physicians are pursuing a theory of mass hysteria in the search for a cause of the mysterious sickness. . . .

Saturday—Today Dr. C. H. Foreman and the experts from the U.S. Public Health Service conferred on the findings of the laboratory tests on several insects taken from the Montana Mills plant to be analyzed at the Communicable Disease Center. . . .

The physicians have advised the company to fill the building with a residual spray that would kill off the bird mite. Officials said they would take this precaution over the weekend. But the investigation continues, say the experts, because nothing specific has been found as yet.

There was one more case of the unknown sickness at the plant today.

Sunday—Nervous disorder, publicity and lastly a bug's bite caused the outbreak of a "very real" and mysterious sickness at Montana Mills. That's the opinion of physicians who carried out extensive investigations. The illness was characterized by nervousness, nausea, weakness, numbness and insect bites.

They summarized it like this: The sickness was definitely real but related to overtones of anxiety and nervousness. They hesitated to use the word "hysteria." The least important factor was the bite of an insect. . . .

Monday—Business was back to normal at Strongsville's Montana Mills plant today. The elusive bug apparently is a thing of the past. And, according to all the experts the rashes and other ailments which caused the trouble can be traced to a bug all right—but a mental one rather than one which crawls or flies. . . .

In any event, as one exterminator put it: "Whatever has been here ain't here now."

A SCHEMATIC OVERVIEW

Montana Mills is a rather large plant employing 965 workers. A subsidiary of a large northern concern, it was relatively new in the area, the sprawling one-story building being only 2 years old. It is an unusual company in that it carries out all operations from spinning raw fibers into thread to the manufacture and distribution of finished women's clothing. Thus, there are several different departments separated in the various parts of the one large building. . . .

The distribution of the workers by sex, department, and shift is significant for our purposes. . . . The plant works three shifts. The vast majority of the workers (728) worked on the first shift (8:30 A.M. to 4:30 P.M.) at the time of the incident. Of those on the first shift, only about one-fifth were men, whereas the majority (59%) of those on the second and third shifts were men. This difference was largely due to the fact that the dressmaking operations were carried out only during the first shift, and these operations were carried out by women. . . .

Finally, it must be noted that June is a month of peak production in this plant. Given the seasonal nature of the basic product, women's clothing, time is of the essence, and June is a crucial month in the production of the fall line. Because of this, much overtime was worked by those in the dressmaking departments. . . . In a business which consistently faced peaks and valleys, then, the epidemic occurred at a peak of production.

It is difficult to specify when the epidemic began. As some of the news releases indicated, workers had complained of insect bites for several weeks before the first group of serious cases brought the situation to the attention of the newsmen. In fact, sections of the plant in the dressmaking departments had been sprayed three times with Malathion the week prior to the outbreak. In spite of this, the complaints continued. Most of the complaints in this earlier period, however, did not lead to medical attention. Somewhat arbitrarily, therefore, we have followed the lead of the medical investigators in using the period from the Friday prior to the Wednesday outbreak through the Monday following that Wednesday as the period of investigation. This is a period of 11 days which includes two weekends.

Within this 11-day period 62 plant employees were known to have been seen by physicians either at the plant or outside. Of these persons, 57 visited the physicians specifically because of bites and associated symptomatology. Most of the other 5 had other assorted complaints which, while not specifically defined as due to insect bites, could be defined as associated with the epidemic. These included nervousness, a burning on the calf, fainting, numbness in an extremity, feeling "like a balloon ready to burst," and an inability to turn the head. Since the epidemic was presumably a function of factors other than insect bites, all of these 62 cases must be considered "affected cases," even though some errors of classification seem almost certain.

Of the 62 cases, 59 were women. . . . Women on the first shift accounted for all but 4 of the cases, and all of these 58 cases were women in the dressmaking departments or in a cloth-mending department adjacent to the dressmaking area. . . .

Of the 62 cases, 50 occurred on the Wednesday and Thursday referred to in the news report. Five occurred on Tuesday, and 4 occurred on Friday of that week. The only earlier case was on the previous Friday, and 1 case occurred each on the Saturday and Monday following the first news release on the epidemic. Thus, within the 11-day period, 95% of the cases reported occurred in 4 consecutive days with 80% of them occurring on 2 consecutive days. . . .

Whatever our approach to this phenomenon, there are thus several basic facts which must be taken into account. The epidemic occurred during the peak production season. It came and went rather quickly, lasting for all practical purposes a week or less. It rose rapidly to a peak and just as rapidly disappeared. Those who were affected were almost all women, and almost all of them worked on the day shift. All but a very few cases were found in one functionally and spatially separate section of the plant. Whatever "it" was, it struck first-shift women in the dressmaking departments more consistently than anyone else. And, most important of all, "it" could not be explained in any "normal" way.

DESIGN AND BASIC FINDINGS

Our investigation was based on a series of interviews with the workers in the plant. Management permitted us to conduct it on two conditions: We could not begin until 2 months after the incident, and we could not tell the workers that the incident was our primary interest. Reference to the incident was made only in passing after most of the interview had been completed.

We restricted the interviews to women on the first shift, and the analysis focused on those white women in the dressmaking departments. We interviewed all of the women known to have been affected and a random one-fourth sample of others in the same departments. There were thus 56 affected cases and 88 nonaffected cases.

During the interview, after alluding to the incident, we asked: "Did anything like this happen to you?" Surprisingly, 17 of our presumed nonaffected cases said "Yes," and they told us about being bitten, feeling ill, etc. Though we had not anticipated this, it required us to make a three-way comparison in the analysis. We referred to our three groups as affecteds, self-defined affecteds, and controls. (The latter is a term we would no longer use, but it is retained here for consistency.) The central focus of our investigation was to differentiate among these three groups of women. Because some were affected and others evidently were not, our most basic task was to differentiate between the controls and the two kinds of affecteds, but because the affecteds had come to the attention of physicians during the incident and the self-defined affecteds had not, we needed to understand that difference as well.

Not surprisingly, 2 months later, those most affected by the incident were most likely to say they believed in "the bug" or some other physical cause of the illness. Although only 51% of the controls gave this response, 71% of the self-defined affecteds and 81% of the affecteds did so.

Our theoretical position led us to focus the analysis on three kinds of factors thought to be relevant in understanding a case of mass psychogenic illness. We thought that such cases occur under conditions in which a number of persons find themselves under stress, so we looked for evidence that those who were affected most were under a heavier strain than those who were not affected. The personality of the individuals presumably also makes a difference. We expected that women who were more nervous and tense, and those who more easily adopted "the sick role" might be more easily affected. Finally, the pattern of spread of the illness, over a period of time, suggested that some interpersonal influence process might be important, so we also focused on the social relations the women had with each other. A brief statement about our findings regarding each of these factors will suffice.

Strain. We investigated various sources of strain, both on and off the job. Since the workers were women, many of whom were married and had children, conflicts between work and family responsibilities (especially with overtime work) seemed likely. We found four measures of strain that differentiated between the affecteds and the controls. Affecteds were more likely to have worked a lot of overtime and to provide half or more of the family income. They were also more likely to see variation in output in their section, thus being more aware of possible invidious comparisons among workers. And, finally, they were less likely to mention their supervisor as one they would go to if they had a problem, an indication, we thought, of felt strain with the supervisor. An index, composed of these four items, clearly differentiated between the affecteds and the controls. However, the self-defined affecteds did not score high on it; they looked very much like the controls, and seemed to be under much less strain than the affecteds.

Personality. Our least expected findings occurred in this part of the analysis. We had expected the affecteds to have high scores on the Cornell Medical Index (acknowledging a large number of physical symptoms) and to more readily adopt the sick role (see a doctor with minor symptoms). Instead, they acknowledged *fewer* physical symptoms than either the controls or the self-defined affecteds, and they did not say they would more readily adopt the sick role. Although these outcomes were, and remain, somewhat puzzling, we adopted a tentative interpretation that involved the concept "denial." We noted that the affecteds not only acknowledged fewer symptoms, on the average, but many more of them than either of the other groups refused to acknowledge *any* symptoms. Although the symptom list offered by us included very minor difficulties, 30% of the affecteds denied having any of them, compared with 20% of the controls and none of the self-defined affecteds. This finding reminded us of another, encountered in the analysis of strain. A number of the women had small children, and we assumed that overtime would present a problem with child care. Some of these women denied that it was a problem and some even said they liked to work overtime. We referred to this as role-conflict denial. Only 10% of the controls and 6% of the self-defined affecteds gave this kind of response, but 29% of the affecteds did so. Finally, we were also reminded that when we asked the women the open-ended question about their involvement in the incident ("Did anything like this happen to you?"), four of the affecteds said "No," even though there was a medical report on their involvement.

We interpreted all of these bits and pieces of evidence as suggesting that at least some of the affected women tended to deny difficulties, and we thought that probably such women would be less likely to cope successfully under conditions of strain. They would thus presumably be more likely to become seriously upset ultimately, exhibiting symptoms that would require medical attention. Interestingly, the self-defined affecteds were quite different. They readily acknowledged symptoms as well as role conflicts—in fact, that was what brought them to our attention during the interviews, and it might have been the reason for their not having become so sick that they had to see a doctor. Although much of this is speculative, based on inadequate evidence, the interpretation is at least worthy of consideration.

Social Relations. The idea of "contagion" is central to most discussions of mass psychogenic illness. Although the term is used in different ways, it clearly suggests that earlier cases have some effect on the probability of a person's being stricken. On the basis of previous literature, we perceived three possible connections between a woman's social position in the factory and the ways in which contagion might occur. It might be that women who were social isolates would be more likely to be stricken because they did not have the social support in the work setting that would help them cope with the threat. It seemed even more likely, though, that women who had social relations with others who had been stricken might themselves fall victim to the contagion, because the illness would seem "closer" and the threat very real if a close associate got sick. Finally, if one interprets contagion as a kind of "crowd response" in which some sort of emotional identification with others in the same setting is sufficient to make one person's experience meaningful to another, social relations might be wholly irrelevant to the path of the contagion.

Actually, we found some evidence in support of all three of these expectations. Overall, the most general finding is that both affected and self-defined affected cases are linked together by bonds of friendship and work-related associations (eat lunch together, car pool members, work partners, etc.). The closer the kind of interpersonal link (e.g., friend rather than work partner), the greater the tendency for the members of each of the three categories to choose each other. But at all levels there are more ties between affecteds and self-defined affecteds than between either of these and the controls. The three categories of women thus all have some characteristics of a "group," but the two groups of affected and self-defined affected women overlap considerably.

However, when the sequence of affected cases is considered, evidence is found for both of the other two patterns. The early affected cases, those that occurred before the two days on which most were stricken, were much more likely to be social isolates than were the other affected women. Thus, those who evidently established the idea that "something is wrong" were outside these group networks. Similarly, those women stricken late on the second big day and afterwards tended to be outside the social networks of the affecteds and self-defined affecteds. However, they were not isolates, because they received friendship and other social tie nominations from other controls. There is some evidence, therefore, that the later contagion spilled out of the social networks of affected cases. (Most of the cases that occurred outside the dressmaking departments also came late in the period of contagion.)

This analysis thus suggests that the flow of the epidemic may pass through several phases. It may begin when a few socially isolated women exhibit a form of behavior that is rather dramatic, so dramatic that it cannot help but be noticed. With the observation of these first few cases, a reasonable explanation evolves that is meaningful in that setting. Both the visibility of the illness and the reasonableness of its explanation increases as the behavior pattern spreads to women who are more socially integrated, and the contagion then follows rapidly along sociometric channels. Finally, though, the cases become so numerous that the threat becomes visible to everyone in the situation and the credibility of the explanation increases. As almost everyone begins to accept the reality of the threat, cases begin to occur throughout the population, and the pattern becomes a kind of "crowd response." Thus, all three theoretical positions we originally considered received some support, though the core of the epidemic seems to have followed a set of sociometric channels.

Whereas the analysis of social strain and personality data had left us in doubt about the place of the self-defined affecteds in the epidemic, the analysis of social relations helped clarify the matter. The earlier analysis showed them to be rather sensible women who were realistic about their situation and were capable of coping with the strains they experienced. It was thus puzzling that they had become affected at all. However, here we find that these women were closely linked in a number of ways with the affecteds. The network of social ties in which they were involved presumably tended to increase the strain they felt and made it more likely that they would experience symptoms. At the same time, they were evidently better able to cope with those symptoms on their own and were thus less likely to come to the attention of the medical authorities.

PERSONAL CHARACTERISTICS, SOCIAL RELATIONS, AND MODE OF RESPONSE

Several of the earlier findings have suggested that *how* an affected woman responded during the epidemic was an important variable, and we had several cues to factors associated with different modes of response. We differentiated among the affected women according to whether they fainted, almost fainted, or did neither. We reasoned that the woman who fainted was evidently overcome with fear and tension before she could get medical assistance, whereas the others evidently got to the doctor before experiencing such overwhelming emotions. . . .

Those who were deniers presumably would tend to delay acknowledging a problem and should be more likely to faint, whereas those who easily acknowledge symptoms should be unlikely to do so. Those who had friends who were also affected might be expected to be alerted to the reality of the danger and be more likely to seek aid before they were overwhelmed. Those who had a generally high inclination to adopt the sick role would presumably seek assistance under conditions of less distress than those who normally had a low inclination to seek medical aid. We examined each of these possibilities. . . .

In this analysis . . . the two factors that stand out are role-conflict denial (which evidently delayed a woman's admitting the need for help and thus increased the probability that she would faint) and social ties to earlier affected cases (which seems to have accelerated the tendency to seek aid and thus lowered the probability of fainting). We were surprised to find that symptom denial and inclination to adopt the sick role were not related to the type of response the women exhibited. We knew from an earlier analysis, however, that both of these variables were related to having ties with earlier affected cases. Those with ties to earlier cases tended to be symptom deniers more often and to have a lower inclination to adopt the sick role. It thus seemed possible that there was some kind of interaction effect between having social relations with earlier affecteds and some of our measures of personal characteristics. To test this possibility, we computed the proportion who fainted for each of the major personal characteristic categories, holding social relations constant. The findings are reported in Table 1.1.

TABLE 1.1
Distribution of Cases of Fainting According to Symptom Denial, Role-Conflict Denial, and Inclination to Adopt the Sick Role, Controlling for the Presence of Social Ties with Earlier Cases[a]

	Affecteds	
Characteristic	*Linked Percentage*	*Not Linked Percentage*
Denial of all symptoms	36.4 (11)*	33.3 (6)
Acknowledge some symptoms	7.1 (14)	56.0 (25)
Role-conflict denial	57.1 (7)	55.6 (9)
No role-conflict denial	5.6 (18)	54.5 (22)
Low inclination to adopt sick role	20.0 (10)	71.4 (7)
High inclination to adopt sick role	13.3 (15)	50.0 (24)

*The number in parentheses following each percentage is the number of affected cases in that category. For instance, there are 11 affecteds who are symptom deniers and who have links with earlier affected cases. Of these, 36.4% fainted.

[a] Kerckhoff and Back (1968). Reprinted by permission.

Although the patterns are somewhat different, in all three cases we find that when the woman had both social relations with earlier affecteds and the particular personal characteristic, she was least likely to faint. The combination of being a nondenier and having a tie with an earlier case or having a high inclination to adopt the sick role and having such a tie produced the fewest fainters. The interaction effect between social relations and the personal characteristic is most apparent, however, for the two measures of denial. . . .

An even more fundamental point to be made on the basis of this analysis is that the role of social ties with affected cases is different depending on whether we are trying to understand *how a woman gets sick* or *how she comes to the attention of the medical authorities*. Having ties with any other affected case, whatever the order of reporting to the doctor, tends to be associated with getting sick. But having ties to a case that has already had medical attention tends to be associated with seeking such attention before fainting. . . . If we use the emotional experience as our dependent attribute, the order of affected cases is not important, but if we use the behavior which follows from that experience (fainting or going to the doctor) the order affected is quite important.

A MULTIDIMENSIONAL MODEL

Throughout our analysis it has been apparent that no one of our variables is a completely successful predictor of a woman's becoming affected, even though the three different types of predictors (strain, personal characteristics, and social relations) have all proved to be effective. In several instances we have seen that combinations of these variables have proved more effective than the individual measures. The implication throughout has been that different kinds of people, under different kinds of strain, and with different patterns of social relations, respond differently. We should thus expect that some kind of multidimensional model would be most effective in organizing our findings.

If we are to attempt an overall statement of the contribution our analysis makes to an understanding of the dynamics of the epidemic, however, we must again recognize that we have to this point been dealing with only a part of the population of plant workers who were exposed to the experience of the epidemic. Although there seems to be good reason to leave the Negroes and those outside the dressmaking departments out of our analysis, it is also true that thus far we have not even dealt with all of those white women in the dressmaking departments. The fact that we have a sample of those who did not report to the doctor plus the total population of those who received medical aid complicates the situation. This is particularly true because the self-defined affecteds have played such an important part in our analysis. Ideally, a summary statement should reflect the process occurring throughout the relevant population (i.e., white women in the dressmaking departments). We must therefore make some assumptions about those women in the population whom we did not interview. The most reasonable assumption appears to be that our sample of nonaffected cases (controls plus self-defined affecteds) is representative of the population of nonaffected cases, and that the distribution of all characteristics within that sample is the same as the distribution within the population. We must also assume that we actually achieved a one-fourth sample of those nonaffected cases. Thus, whenever we examine the distribution of a particular characteristic, we need to multiply the data we have on the nonaffected cases by four and then add it to the data we have on the affected cases. This should then allow us to make the best estimate of the distribution of these characteristics within the total white female population in the dressmaking departments. Given such assumptions, we can analyze the distribution of the important characteristics within what can most legitimately be called the "population at risk."

In order to present in summary fashion the effect of the kinds of factors we have discussed earlier within the total population at risk, we selected the three most important measures from the earlier analysis and combined them. First there was the summary measure of strain which we viewed as an index of the situations in which our subjects found themselves. . . . This was then combined with the measure of the woman's position in the interpersonal influence process during the epidemic. Since we were going to use as our dependent measure the fact that a woman was or was not an affected case, we counted intimate ties with *any* other affected woman as an indication of a woman's social position. Such ties ranged in number from zero to three. Finally, within the categories defined by these situational and sociometric measures, we differentiated between two kinds of persons, those who were and were not deniers by one or both of our measures. Since there were 5 possible levels of strain, 4 degrees of social linkage with affecteds, and 2 types of persons, this led to 40 possible combinations. Because some of the combinations did not appear and others were very infrequent, we reduced the number of categories by collapsing the strain and social position measures into 3 levels each. This provided the basis for an 18-cell table.

The data are reported in Table 1.2. The first number in each cell is the number of affecteds of that type, the second is the number of self-defined affecteds, and the third is the number of controls. Both of these latter two numbers were arrived at by multiplying the number of self-defined affecteds and controls of each type in our sample by four, thereby providing an estimate of the number of such women within the population at risk. The total number of cases in the table is 408.

If we wish to "predict" which cases were affected, using the three factors of strain, social influence, and denial, we might pick out cells containing 56 cases in such a way as to maximize the accuracy of prediction. The cells with the double asterisks contain a total of 54 cases. To fill out the required number we might draw randomly 2 more cases from some adjacent cell. Since no such cell has half or more affecteds, we may assume that both cases so drawn would be

TABLE 1.2
Distribution of the Population at Risk Classified by Level of Strain, Ties with Affected Cases, and Denial[a]

		Level of Strain		
		0-1	*2*	*3-4*
No Affected Intimates	No Denial	4-16-96	2-4-36	2-0-0**
	Denial	3-0-56	2-4-4*	3-0-0**
One Affected Intimate	No Denial	4-20-52	2-8-12*	4-0-4**
	Denial	3-0-4*	7-0-4**	4-0-4**
Two or More Affected Intimates	No Denial	3-4-0*	3-4-8*	2-8-0**
	Denial	3-0-0**	2-0-4**	3-0-0**

NOTE: The first number in each cell is the number of affecteds, the second is the number of self-defined affecteds, and the third is the number of controls. An "intimate" is a friend, one to whom one would go for advice or would expect to come to her for advice, a carpool or lunch partner, or one defined as the best worker in the section. "Denial" includes both the denial of symptoms and role-conflict denial. Cells with double asterisks designate those categories in which more affected cases are found. Those with single asterisks, together with those with double asterisks, designate those categories in which more affected and/or self-defined affected cases are found (see text).

[a] Kerckhoff and Back (1968). Reprinted by permission.

nonaffecteds. How well does such a table order the cases? Perhaps most important is the fact that there is a simple pattern to be seen. First, there is a general tendency for the proportion of affecteds in the cells to increase as we move from upper left to lower right. All but one of the cells with double asterisks are in the last column or last row of the table. Second, either an extreme level of strain or close association with two or more other affecteds tends to lead a woman to be affected, irrespective of the other factors. Third, denial, when added to either social influence or strain, greatly increases the probability of being affected.

Also, it is striking that so many of the 56 cases we have just singled out are affecteds. Not only are 30 (or 53.6%) of them affecteds, but of the remaining 352 cases, 326 (or 92.6%) are *not* affecteds. There is thus an overall accuracy of designation of 87.3%. This level of accuracy, of course, is far greater than could be accomplished using any of the three kinds of measures individually. Although it is clear that the strain and social relations measures predict well in the more extreme cases, the *combination* of variables is clearly important at the less extreme levels.

We may also ask whether such a table helps to single out the self-defined affecteds as well. Because there were 17 such cases in our sample, we can assume that there were 68 in the population at risk. These, plus the 56 affecteds, would give a total of 124 women who experienced symptoms during the epidemic. The cells with the single and double asterisks contain 115 cases. To fill out the required number, we could draw 9 cases randomly from the cell in the first row and second column, and we could assume that 1 of these 9 cases would be either affected or self-defined affected. Of these 124 cases, 72 (or 58.1%) are either affecteds or self-defined affecteds, and of the remaining 284 cases 232 (or 81.7%) are controls. There is thus an overall accuracy of designation of 74.5 percent.

The important part of these results, however, is the high proportion of affected (or self-defined affected) cases in the critical cells, since only a very small percentage of the total of 408 were either affected (13.8%) or self-defined affected (16.7%). Perhaps the best way to illustrate the effectiveness of the three independent attributes is to note that the cells in Table 1.2 which are marked with asterisks represent less than three-tenths (28.1%) of the population at risk, but they contain more than three-fourths (76.8%) of those who received medical treatment and more than half (57.3%) of all those who experienced symptoms. The three independent measures are thus much more effective in discriminating between those who did and did not go to the doctor than they are in discriminating between those who did and did not experience symptoms, although they do both to a significant extent.

A GENERAL OVERVIEW

Even though our data fail to provide a completely adequate explanation of the facts of the epidemic, the analysis presented . . . has indicated that there is more order in this seemingly chaotic situation than one might at first have expected. . . . Therefore, it may be well to state in a more succinct summary form what we believe this analysis has done to clarify the order in the chaos. We will do this through a set of statements. . . .

Most affecteds were exposed to a great deal of strain, or were deniers, or both. . . .

The difference between affecteds and self-defined affecteds was largely due to differences in personal qualities and in sources of strain. . . .

The difference between self-defined affecteds and controls was largely a difference in social relations and personal qualities. . . .

If the strain experienced was sufficiently great, it alone was likely to lead a woman to become affected during the epidemic. . . .

If social influence was strong enough, it alone was likely to lead a woman to become affected. . . .

Among those affecteds with social ties with other affecteds, the ones who became affected first exhibited the most extreme symptoms. . . .

The combination of denial and a moderate level of strain or social influence greatly increased the tendency to become affected. . . .

Personal and social characteristics interact to affect the tendency to exhibit the most extreme symptoms. . . .

There were many persons (the self-defined affecteds) who played an important part in the development and spread of the contagion who were not known by the authorities to have been affected. . . .

CONCLUDING OBSERVATIONS

. . . To sketch the various patterns that need to be combined it is necessary to broaden our perspective to include at least the whole plant as well as a more extended period of time than we have investigated. This is especially true for an understanding of the central belief in the case, the belief in a poisonous insect. Some of the early news items alluded to the fact that there had been complaints of insects for several weeks before the epidemic. Evidently during this time there had diffused rather widely in the population of the plant the belief that there was an uncommon number of insects and that they constituted at least a nuisance. It is within this context that the first few cases must be seen. As the symptoms and the knowledge of sickness spread, the link between this earlier belief and the immediate experience was made. The insect changed from a nuisance to a threat.

We must assume that in the face of this threat a number of different things began to happen, the most obvious and most central from our perspective being the rapid increase in reported symptoms. But there must have been other patterns of contagion going on at the same time. Undoubtedly in the face of this threat some of the women (with or without having experienced symptoms) evolved the solution of leaving the plant and there would presumably have been a contagious adoption of that solution. Others probably turned to officials in the plant for remedial action, and there would presumably have been a contagious move in that direction by some of the workers.

The increase of reported symptoms evidently understates the actual increase in symptoms—our self-defined affecteds attest to this. Given the experience of symptoms, a woman might or might not have become known to the medical authorities. She could have "clocked out of there," taken an aspirin, "talked it out" with a friend or her supervisor, *or* gone to the doctor. Except for the accidental discovery of the self-defined affecteds, our concern would have been exclusively with those known to the medical authorities. Even within that group, however, it proved useful to differentiate between those who went to the doctor on their own and those who fainted before they could find aid. A relevant aspect of this differentiation is the suggestion that there developed during the period of the epidemic a greater tendency to seek aid before being overcome. Thus, for those who did develop symptoms and could not cope with them in any other way, there evidently evolved an acceptable solution where one had not existed before. There was evidently a process of social facilitation in the use of this method of coping. In addition, there was also evidence that persons with milder symptoms were likely to report to the doctor in the latter part of the epidemic. This could be interpreted either as an indication of a heightened awareness of physiological disturbances and thus a tendency to define mild symptoms as significant, as an indication of a desire to cooperate with the officials by keeping them informed of even mild "attacks," or as a means of using a socially approved way of accomplishing some other goal (getting attention, going home, etc.).

This seems to mean that in any such case there are several processes of dissemination going on at the same time, all of which are interrelated but any one or combination of which may become the focus of attention: (1) the spread of a belief in a threat; (2) the spread of the experience of symptoms; (3) the spread of relatively unobtrusive methods of coping with the threat and/or the symptoms; (4) the spread of cases of collapse in the face of the threat and/or the experience of symptoms; (5) the spread of the solution to the experience of symptoms via seeking medical aid. Only the last three are likely to bring the process to the attention of outsiders, and, depending on the "unobtrusive" methods devised, it may be that only the last two will do so. All of these processes are to be expected whether the "cause" of the symptoms is "real" or "purely imaginary." In fact, we put these terms in quotation marks because we suspect that it is never possible to use them in such cases with complete confidence, and the social effects will be the same in any event.

REFERENCES

Kerckhoff, A. C., & Back, K. W. The June bug: A study of hysterical contagion. Englewood Cliffs, N.J.: *Prentice-Hall, Inc.*, 1968, pp. 3–18, 130–156, 180–200. (Excerptation by permission of *Prentice-Hall, Inc.*)

2 Outbreaks of Mass Hysteria at Workplaces in Singapore: Some Patterns and Modes of Presentation

W. H. Phoon
Ministry of Labour, Singapore

ABSTRACT

Although outbreaks of mass hysteria have been commonly described in schools, in recent years, this phenomenon has been noted to be not uncommon in workplaces. This paper summarises 6 such outbreaks in the island Republic of Singapore and discusses some features. The outbreaks usually occurred in electronics factories which were modern, clean, and air-conditioned. The affected persons were mainly female workers belonging to one racial community in cosmopolitan Singapore. Certain cultural factors might be contributory, e.g., belief in spirit-possession. The marked "infectiousness" of mass hysteria is emphasized, as it was noted to spread very rapidly, and even from one factory to another. Certain persons were more prone to the phenomenon, and had repeated attacks. There were essentially 2 types of presentation—the violent and the frightened. Much more study needs to be done into this multifactorial phenomenon before it can be adequately prevented.

Cross-cultural studies examining outbreaks of mass hysteria or "epidemic" hysteria have typically reported cases in school settings in various countries (Ebrahim, 1968; Knight, Friedman, & Sulianti, 1965; Tan, 1963). But in recent years, this phenomenon has been noted to occur not infrequently in factories and other places of work, where its effects and implications may be much more serious in terms of economic considerations and work morale. This paper summarizes six outbreaks of what may be termed "occupational" mass hysteria in the island Republic of Singapore, from 1973 to 1978, and discusses certain interesting features.

Case Study 1

Three factories, situated near each other in a light industrial estate comprising blocks of flatted factories, were affected in the first series of outbreaks. The first, and the main factory affected, was involved in assembling television sets. It employed 802 female factory workers (of whom about 65% were ethnically Malay, the rest being mainly Chinese and Indians), and 97 male workers, working 2 shifts, from 7 a.m. to 3 p.m., and from 3 p.m. to 11 p.m. On Saturday, the 13th of January, 1973, at about 7 p.m., a female worker suddenly started screaming and then fainted. When the supervisor and some of her colleagues carried her to the clinic on another floor, she started screaming again and struggled with them. Subsequently other women were affected. Because of these incidents, the factory closed earlier that night.

The next 2 days were holidays, being a Sunday and a public holiday. As all the women affected were Malays, the management engaged a *bomoh* or native medicine man to exorcise any evil spirits as it was believed that the incidents were a manifestation of spirit possession. But when the factory reopened on Tuesday, more women were affected beginning at 2:30 p.m.

On January 17th, the hysteria outbreak spread to a nearby factory manufacturing electrolytic capacitors, which also employed a predominantly female labor force. A few days later, a third factory reported one hysterical case.

The whole outbreak lasted from January 13 to 27, and a total of 97 persons were affected, giving altogether 107 "spells" (or attacks), some having repeat attacks. See Table 2.1 (Chew, Phoon, and Mae-Lim, 1976.)

It was noted that 3 of the workers affected on the first day, had repeat attacks on the second day when the factory reopened after 2 days of public holidays. All repeat attacks involved females, except one Malay male, who experienced only fear. Except for one Indian, all the cases were Malays. There were 3 main modes of presentation of the cases:

1. Hysterical Seizures: These were characterized by screaming, struggling fiercely and general violence. A very violent woman would require 4 men to hold

TABLE 2.1
Number of Spells and Subjects

Factory	*Hysteria*	*Frightened*	*Total*	*No. of Subjects*
A	39	51	90	84
B	16	0	16	12
C	1	0	1	1
Total	56	51	107	97

her down. (Intramuscular injection of chlorpromazine or other tranquilizers, e.g., valium [diazepam] did not seem effective.) The affected worker would ignore questions or attempts at reassurances and might even scream more. Some would scream that they saw a dark figure coming to strangle them. Some would laugh between screams.

2. *Trance States:* This was a "subgroup" of the hysterical presentation. While being violent, or before becoming violent, a worker would suddenly speak as though she was someone else, and claim that a "jin" or spirit was speaking through her. Four of the hysterical cases developed such a state, which lasted only a few minutes.

3. *Frightened Spells:* The affected workers only expressed a feeling of unexplained fear, and might complain of feeling cold, numb, or dizzy.

Case Study 2

The second major outbreak involved a food factory. The factory produced dried cuttlefish. It occupied a 4-story building with an employment of about 350, of whom 300 were females, and about 70% were Malays. Working hours were from 7 a.m. to 5 p.m. (with staggered starting and finishing time), Monday to Friday, with work on alternate Saturdays. There was no shift or night work.

The outbreak started on Saturday, September 29, 1973, when a female worker suddenly started screaming and crying. She then became violent, requiring several men to restrain her. This was followed by two other women getting the attack. It was said that the women saw a black person with hairy arms trying to frighten them.

The next day being a Sunday, the factory was closed. Nothing happened on the next 3 working days until Thursday, October 4th, when cases of hysteria started at 8 a.m. A total of 10 women were affected that morning, including 3 who had been affected on the first day. As there was great confusion, the management closed the factory in the afternoon.

A *bomoh* or medicine-man was called to exorcise evil spirits. A repeat exorcism was carried out a few days later, as cases of hysteria continued to occur. A total of 56 "spells" (17 violent and 39 frightened) involving 28 persons occurred between September 29 and October 15, 1973.

Those affected were all Malay females, with two Malay males. (One of the Malay males were violent, but the other only experienced fear.) A feature of this outbreak was the high rate of recurrence (see Table 2.2). It was interesting to note that all 3 of the women affected on the first day of the outbreak had repeat violent attacks on the second day when cases occurred again, although there were 3 working days in between with no attack.

TABLE 2.2
Recurrent Attacks (Spells)
in those Affected

No. of Attacks (Spells) Per Person	*No. of Persons*
1	12
2	10
3	2
4	2
5	2
Total	28

Case Study 3

Another series of episodes involved two adjacent factories. The first manufactured dry cadmium-nickel batteries for use in cameras, calculators, etc. It occupied 4 floors of a modern factory building, and employed about 500 workers, mostly young females in their teens and twenties, about 70% of whom were Malays. There was generally only one shift, from 7:30 a.m. to 5:00 p.m., Monday to Friday, with occasional overtime on Saturdays on a voluntary basis.

On Friday, July 23, 1976, at about 2:40 p.m., a female worker in the Final Assembly Section suddenly screamed and collapsed to the floor, and lay there quietly. A few workers helped carry her to a rest area in the canteen on another floor. On the way down, she started screaming and became violent. About 10 minutes later, another female worker started to scream. Soon afterwards, others followed. Because of this, that section of the factory was closed early, and overtime work scheduled for the next day (a Saturday) was called off.

On the following Monday, at about 3:10 p.m. hysterical cases recurred. Altogether, of 6 workers who exhibited violent behavior, 4 were repeats, who had had similar attacks on the previous Friday. Six additional workers were frightened.

A total of 28 spells occurred in 18 persons, between July 23 to 30. A summary of the cases is presented in Table 2.3.

All those affected were females, 16 of the 18 were of Malay origin, one Chinese, and one Pakistani. The latter two experienced only fear.

On July 29, 1976, cases appeared in a neighboring factory manufacturing cameras, and employing 3800 persons, 50% of whom were Malays. A female Malay worker complained of a headache, and suddenly fell down, screaming. (When interviewed after the incident, she denied seeing any spirit or figure, but said she had felt frightened.) Three female operators and a male supervisor then helped her to the office to rest. Two of the female operators, while in the office, also experienced fear, and started to cry and scream.

TABLE 2.3
Summary of Cases in Outbreak (R = Repeat)

	No. of Spells		
Date of Occurrence	*Hysterical*	*Frightened*	*Total*
Friday (7/23/76)	6	1	7
Monday (7/26/76)	6 (4R)	6 (1R)	12 (5R)
Tuesday (7/27/76)	0	2	2
Wednesday (7/28/76)	0	1 (1R)	1 (1R)
Thursday (7/29/76)	2 (2R)	2 (1R)	4 (3R)
Friday (7/30/76)	0	2 (1R)	2 (1R)
Total	14 (6R)	14 (4R)	28 (10R)

The outbreaks in this factory lasted from July 29 to August 4, 1976. There were a total of 30 spells with 20 hysterical seizures and 10 frightened occurrences, affecting 23 persons of whom 7 had more than one spell each. All except one were females. The sole male involved only had a feeling of fear. Of the 23 affected, 22 were Malays, and one an Indian.

This episode also illustrates the great likelihood of recurrence of an attack in an affected person, especially among the first few cases. The first person to be affected on the first day was also the first case on the second day, after the weekend. Out of the 6 hysterical cases on the second day, 4 were repeats of those affected on the first day. In this outbreak, urine samples from a few of the hysterical women were taken for analysis for morphine, codeine, and methaqualone. None was detected.

Case Study 4

The next hysteria outbreak affected not a factory, but a large company engaged in providing telecommunications services. The outbreak occurred in a branch telephone exchange with 527 persons, all females, 203 or 38.5% being Malays. They worked shifts.

On October 20, 1977, at about 4:30 p.m., one of the Malay telephone operators fainted, and suddenly started to scream. Similar behaviors followed a little later among a number of other women. Altogether 17 women were affected between 4:30 and 7:00 p.m. A doctor was called who gave injections of tranquilizers and called the ambulance to take the affected women to the hospital. A number of cases became violent and had to be physically restrained. Altogether, between October 20 and November 8, twenty-two operators were affected. Ten of the women exhibited violent behaviors, ten reported feelings of fear, and two were both violent and frightened. Thirteen were Malays, 5 Chinese, 2 Indians and 2 of other races.

In 1978, there continued to be sporadic attacks among some of the telephone operators affected in the outbreak in October/November, 1977. One of the 2 women who had 2 attacks in that outbreak, had her second attack on November 8, 1977 at another branch telephone exchange to which she was transferred a few days after she had her first attack. She was subsequently transferred again to another exchange where she had her third attack on April 20, 1978. On October 26, 1978, she had her fourth attack at yet another branch exchange to which she was transferred. The other woman who had 2 attacks in October/November, 1977, had her third attack on July 4, 1978 at another telephone exchange. A third woman, after having an attack in October/November, 1977, was later transferred to another exchange, where on April 18, 1978, she had her second attack, after having been told that morning that she would be transferred again to another exchange (which was far from her home). On May 18, 1978, while still at her previous place of work, she had her third attack. (She was at that time, in the third month of her first pregnancy.)

Case Study 5

Yet another episode involved a factory engaged in the manufacturing of ballbearings. It occupied a modern single-story building, and employed 800 workers, 50% of whom were Malays. The incident occurred in one section of the factory, called the Clean Room where 107 female workers, working shifts, were engaged in "sealing" the ballbearings. Adjoining the Clean Room was a Washing Room where ballbearings were degreased in ultrasonic tanks using freon.

On Sunday, July 30, 1978, there were about 20 female workers working overtime in the Clean Room, commencing work at 8 a.m. At 9 a.m. some workers complained of an "acidic smell." At 9:15 a.m., a worker (A) complained of dizziness, nausea, tearing, and a blocked nose. She then visited the toilet, and while at the toilet she fell down. On return to the Clean Room, she fainted. At about the same time, another worker (B) also had similar symptoms. She visited the toilet, and when she returned to the Clean Room, she saw that A had fainted. With the help of two other workers, she carried her to the office. She then also fainted.

A third worker (C) had symptoms of headache, nausea, and tearing. When (A) fainted, all the workers in the Clean Room were evacuated, at about 9:30 a.m. On reaching the reception hall, (C) also fainted. A fourth worker, together with the rest of the workers, left the Clean Room. On reaching the reception hall, she also became ill.

The Company doctor was called. He found all 4 women conscious, but "agitated." One of them (A), became very hysterical, crying, screaming, hyperventilating, and developing carpal spasms. The doctor gave all 4 injections of valium (diazepam) and sent them to the hospital. Three of them were warded for observation. On examination, nothing abnormal was found, but (A) was very "upset" and crying. The next day, (A) developed hysterical hyperventilation, which subsided with an injection of valium.

The Management's explanation for the incident was that one of the workers (in another section) had forgotten to switch on the cooling condensing coils of the ultransconic tanks in the Washing Room. Because of this, the vapor arising from the volatile freon was excessive. The gas was removed by the local exhaust ventilation system at the mouth of the tanks, but as it travelled along the ducting, because of the excessive vapour "overload," some of it leaked out of the system. As the ducting passed through the toilets in the women's restroom, the women must have been affected when they went there.

This explanation was initially thought plausible, but interviews with the 4 affected workers revealed that they developed symptoms while in the Clean Room, and 2 of them did not even visit the toilet that morning. If, on the other hand, there had been leakage of freon vapor from the Washing Room into the Clean Room, (as a complaint of "acidic smell" was made), then it was rather surprising that the other women in the room did not seem affected. The seating arrangement of the employees was such that the 4 affected workers were seated in various parts of the room. Also, the hysterical presentation was rather suspicious. This suspicion, however, was only strengthened when another similar incident occurred in December 1978.

Case Study 6

On Monday, December 18, 1978, soon after reporting for the afternoon shift at 3 p.m., a few workers complained of a "chemical smell" in the Clean Room. At about 3:30 p.m., the supervisor saw two Malay women seated at their tables, with their heads in their hands. They complained of giddiness. They were asked to go to the locker room just outside to rest a while. A little later, a Malay man, in charge of the Washing Room, reported that the two women had apparently fainted in the locker room and he asked the supervisor to evacuate all the workers from the Clean Room. There was a lot of commotion as the employees left the room. Outside the room, some female workers fainted, and were carried to the Conference Room to recover. Oxygen was administered.

There were a total of eight women, all Malays, being resuscitated in the Conference Room, including the first two workers who had fainted in the locker room. The doctor was called. Two of the women become violently hysterical, screaming and struggling, and had to be physically restrained. (One of these was (A), who was also the first worker who had complained of giddiness and later fainted in the locker room.) After injecting them with valium (diazepam), the doctor called the ambulance to send them to the hospital. Seven were warded for observation, while the eighth (A) was observed for an hour and sent home with one day's sick leave. The doctors found nothing abnormal on physical examination of the 7 workers, although they complained of dyspnoea and a "suffocating'' sensation. Chest X-rays, and blood gases done on a few of them, were also normal. All were discharged the next day.

Of the eight women affected, three of them had been affected in the earlier incident in July. The first victim in the December incident was not only among the four affected earlier, but she was also one of the two persons who became ''violently'' hysterical. In the hospital the next day, she also developed ''hysterical hyperventilation.''

GENERAL PATTERNS OF MASS HYSTERIA CASES

The outbreaks of mass hysteria described presented, with some exceptions, a general pattern:

1. A sudden onset (without warning).
2. A high degree of ''infectiousness'' or communicability, spreading not only from person to person, but also from one department to another, and even from one factory to another.
3. A high rate of ''repeats,'' i.e., the tendency for a person, after having experienced one attack, to get another.
4. A likelihood for those affected on the first day of an outbreak to be involved in other attacks on the second and subsequent days.
5. Although a psychological phenomenon, psychological countermeasures did not prove effective, e.g., the calling in of a *bomoh* or medicine-man to exorcise spirits.
6. An outbreak would subside on its own, usually within a week or so.

There were 2 main modes of presentation of cases:

a. The violently hysterical—presenting with screaming, crying and struggling. The violence might be aggravated by attempts to restrain them. During such attacks, they were usually incoherent and ignored questions and attempts to calm them. Injection of transquilizers did not seem particularly effective. They might or might not have visual hallucinations.
b. The frightened—presenting with an unexplained feeling of fear. They might also complain of feeling cold, numb or giddy.

In addition to the above, another presentation, which is more a ''subgroup'' of the violently hysterical, was that of a ''trance state,'' during which the person acted as though she were somebody else, with a spirit of *jin* speaking through her. This presentation, lasting only a few minutes, was uncommon, and was observed only during the January 1973 outbreak.

This interesting phenomenon could be partly explained by the cultural background and traditional belief in spirits by the Malays. According to this belief, living things have a "vital force" which they call *semangat* which plays an important part in maintaining health. Its presence in the body is believed to exert a force externally, preventing evil spirits (*hantus*) from acting on the body. The *hantus* are believed to be ubiquitous but invisible, and it is only when the force of the *semangat* is weakened, that the spirits can gain entry into the body. Sudden fright or shock is believed to drive the *semangat* out of the body, and hence render the person vulnerable to the evil spirits. Females are believed to have weaker *semangats* than men. (Children with fits are thought to be possessed by a *hantu*. When a child's eyes roll upwards during a fit, he is believed to be looking at a *hantu* on the ceiling about to attack him.) Malays generally also believe in *bomohs* some of whom have the *Jin* (or genie). These *bomohs* are able to make charms for good or evil (Kass, 1961).

In spite of such beliefs, however, it was noted in several outbreaks, that the calling in of a *bomoh* to exorcise evil spirits did not stop the outbreak, which subsided on its own. Although an outbreak usually lasted about a week, with daily recurrences, there may be a lapse of a few days with no attacks, and this may give rise to a "false sense of security." Thus, vigilance must be maintained for at least a few days after the last case has occurred.

All the outbreaks occurred in large modern factories or work places. The general work environment was clean, airconditioned, with good lighting and ventilation. In most cases, little or no chemicals were present or handled. It thus seemed that the physical work environment might not have been a factor in such outbreaks. Shift and night work also did not seem to be a factor as outbreaks occurred on all shifts both day and night. It was noted, however, that all the factories involved "assembly line" type of work, which was monotonous and repetitive. Monotony arising from underarousal can be a stress factor (Dixon, 1977; ILO, 1972, 1976, 1978). Some factories in the U.S.S.R. and Sweden have, in fact, introduced modifications in the assembly line type of work, in an attempt to reduce the boredom of repetitive work.

Almost all of the cases involved females, but the phenomenon was not confined to the young or any particular age group, as the age distribution of the affected workers reflected the age composition of the general worker population in those factories. (It was perhaps fortunate that few men were affected, as those affected seemed to possess extraordinary strength!)

Most of the persons affected were Malays. However, in a few outbreaks, statistical evaluation of the racial composition did not seem to find this a significant factor. But as the numbers involved in each outbreak were relatively small, this must be taken with caution. It was noted that such outbreaks usually occurred in factories which had a predominantly Malay worker population.

In general, workers who were affected on the first day of an outbreak were more likely to get repeat attacks on the second and subsequent days. It should therefore be considered in any outbreak, to remove those affected (especially on the first day) and keep them away from the workplace for at least a week (or until the outbreak has subsided). This is important in view of the fact that one attack can so easily trigger an attack in another person. Thus persons involved in one attack should be considered "highly infectious," to be strictly segregated and isolated (Chew, 1978).

To find a "triggering" event that initiated an outbreak was very difficult. In several outbreaks, there were, as far as we could ascertain, no recent change in management or supervisory personnel, or in conditions or arrangements of work. In the 1978 outbreak in the ballbearing factory, however, several of the affected workers, when interviewed discreetly, confided some complaints about the management, and excessive work pressure, including excessive overtime. These complaints were greatest in the Clean Room, where the outbreak occurred. This particular section had a higher labor turnover than the rest of the factory. This set of circumstances may partially explain why the cases occurred only in that particular part of the plant and did not spread to the rest of the factory, as in other outbreaks.

The presentation of cases in the ballbearing factory was also rather interesting and unusual. Unlike the usual presentation of sudden screaming and fainting, the affected workers complained of headache, giddiness, etc. and then fainted. This had initially been attributed to leakage of freon vapor, and it seemed a likely explanation. However, in retrospect, after 2 episodes of a similar nature had occurred, affecting the same workers, whereas other workers were unaffected on all occasions, it would seem unlikely that freon exposure alone could account for all illness occurrences. Clinical and laboratory investigations at the hospital were negative. It is thus important to bear in mind that an outbreak of mass hysteria may mimic some other condition (and vice versa), and it may be in retrospect, in light of accumulated circumstantial evidence, that the actual cause is discovered.

The outbreak at the telecommunications workplace involving telephone operators was also interesting. Recurrence of attacks in a few of these persons occurred as isolated incidents and did not spread to other people. Frequent transfers and change of workplaces seemed to be a factor in these attacks. A subsequent study (Chan, Zee, & Wong, 1979) found that the telephone operators affected in the 1977 outbreak, had a slightly higher Neuroticism score than the controls, but this was not statistically significant. They also had a higher Extrovertion score and a lower Intelligence score, but $p > .05$.

Having looked at the patterns of various mass hysteria outbreaks, we would next have to identify what factors would "trigger off" an outbreak, and what factors would make certain persons more "susceptible" to an attack than others. With regards to the latter, besides studying such demographic data as age, sex, race, religion, duration of employment, educational levels, etc., possible areas of study would include the psychological makeup and personality, intelligence level and level of job satisfaction.

An explanation for such outbreaks affecting mainly Malay female workers has been advanced by Ackerman (1978). The traditional upbringing of Malay females is very conservative. Thus the social norm expected of them is to be submissive, obedient and "well controlled." Self-expression is not a characteristic to be encouraged, it being preferred that girls remained *malu* (i.e., shy and retiring). Malay females, especially if from the rural areas, may encounter psychological stress in adapting to an industrial setting. Added to this, they may also have dissatisfaction over certain work conditions and styles of management (e.g., strict discipline). But because of their upbringing, it is against their nature to openly rebel. Direct confrontation with authorities would bring about social disapproval. It is also possible that monotonous work would leave them much time to "brood over" any unhappiness they may experience.

Traditional Malay culture includes strong beliefs in spirits and spirit-possession. It is thus postulated that, because of internal conflict generated by unhappiness and the inability to self-express or vocalize their unhappiness, for fear of social disapproval, the Malay women react by mass hysteria, i.e., it is used as a "safety valve." Because of the traditional belief in spirit possession, such behavior, far from being condemned, is accepted without any social disapproval. On the contrary, it generates concern and sympathy on the part of parents, colleagues and perhaps even management. However, much more study is required into the various aspects of the phenomenon.

REFERENCES

Ackerman, S. E. *Industrial conflicts in Malaysia: A case study of rural Malay female workers.* 1978, unpublished manuscript.

Chan, O. Y., Zee, K. O., & Wong, C. K. High risk factors in Singapore factory workers. *Occupational Health and Safety,* 1979, *48,* 58–60.

Chew, P. K. How to handle hysterical factory workers. *Occupational Health and Safety,* 1978, *47*(2), 50–54.

Chew, P. K., Phoon, W. H., & Mae-Lim, H. A. Epidemic hysteria among some factory workers in Singapore. *Singapore Medical Journal,* 1976, *17*(1), 10–15.

Dixon, W. M. Occupational mental health in the UK. *Journal of the Society of Occupational Medicine,* 1977, *27,* 143–147.

Ebrahim, C. J. Mass hysteria in school children. *Clinical Paeditrics,* 1968, *7,* 437–438.

ILO. *Occupational health and safety,* 1972, *2,* 1359.

ILO Information. *Swedes lead trend to industrial democracy,* 1976, *12* (6), 5.

ILO Social and Labor Bulletin. *Measures to fight the monotony of assembly line work in a car factory,* 1978, 157–159.

Kass, R. *Some Malay customs, beliefs and practices, and their effects on health and happiness.* (A dissertation submitted for D.P.H.), University of Singapore, 1961.

Knight, J. A., Friedman, T. I., & Sulianti, J. Epidemic hysteria: A field study. *American Journal of Public Health,* 1965, *55*(5), 858–865.

Tan, E. S. Epidemic hysteria. *Medical Journal of Malaysia,* 1963, *18*(2), 72–76.

3 A Review of Mass Psychogenic Illness in Work Settings

Michael J. Colligan
Lawrence R. Murphy
National Institute for Occupational Safety and Health

ABSTRACT

Published and unpublished reports of mass psychogenic illness, defined as the collective occurrence of physical symptoms and related beliefs among two or more persons in the absence of an identifiable pathogen, which have occurred in work settings are reviewed. A number of factors (e.g., job stress, sex-role identification, labor/management relations, physical stress) are identified as potential precipitating conditions and the contagion of symptoms is discussed in terms of the social comparison process as described by Festinger (1954) and Schachter and Singer (1962).

In July 1974, a team of psychologists from the Applied Psychology and Ergonomics Branch (APEB) of the National Institute for Occupational Safety and Health (NIOSH) was asked to investigate a mass outbreak of illness in a garment manufacturing plant in the southwest. This unusual request for psychological evaluation came about after testing for physical agents had failed to detect significant levels of toxicants, although the circumstances of the outbreak had initially implicated the presence of a toxic agent. Within a period of a week approximately 100 women in a plant population of 340 workers (325 women, 15 men) had expressed or displayed symptoms of nausea, dizziness, fainting, and burning sensations in the eyes and throat. A majority of the affected workers reported smelling a peculiar odor or gas in the work environment, which they felt produced their illness. Extensive environmental sampling by industrial hygienists and medical examinations, however, failed to identify the presence of any toxicants capable of producing the observed symptoms. It was felt that the illness may have involved psychogenic components, e.g., stress or anxiety (Shepard & Kroes, 1975).

The immediacy of the situation did not allow for the development of a systematic research protocol. Nevertheless, it was felt that sufficient information could be obtained through observation of the work environment and interviews with affected and nonaffected workers to permit generalizations to other reported cases of this nature. However, a post-investigation literature review was extremely disappointing. Only two published cases of apparent mass psychogenic illness in American work settings (Kerchkoff & Back, 1968; Stahl & Lebedun, 1974) were found. And, as was also true of the NIOSH investigation, the spontaneity of the phenomenon and the absence of a guiding theoretical framework hampered the collection of systematic and comparable data in these two cases. Consequently, the understanding of contagious psychogenic illness is presently more impressionistic than empirical and is limited by the scarcity of existing data.

The initial reaction of NIOSH to the literature review was to conclude that although the phenomenon was extremely interesting and dramatic, it was simply too rare to warrant a systematic research effort. However, over a period of 2 years, subsequent contacts with industrial health professionals, labor and management representatives, and state health officials, indicated that the actual incidence of such illness was much greater than had been suspected. Typically these cases were reported to local authorities and state health agencies who dispatched hygienists and medical professionals to the worksite. Following inconclusive environmental and medical testing, the results were described in internal staff reports that were not made available as general public information. In some cases a suggestion (based primarily on the absence of physical evidence) was made that the illness may have been psychosomatic in nature. No further attempts were made to identify the attributes of the affected workers or the psychological characteristics of the work environment which may have precipitated the outbreak. Unfortunately, merely labeling the phenomenon as "psychosomatic" or "anxiety induced" does little in promoting an understanding of the dynamics of the illness.

It became apparent that a critical need existed for systematic research into the etiology and methods for remediation of mass psychogenic disturbances in industry. In 1976, therefore, the Applied Psychology and Ergonomics Branch of NIOSH initiated a programmatic effort to develop, field test, and refine a research protocol for investigating cases of this nature. This approach involved combining individual interviews with questionnaire data collected from a random sample of affected and nonaffected workers at each worksite. In addition to sociodemographic (e.g., age, sex, level of education, marital and parental status, etc.) and epidemiological information (date and time of illness, symptomatology, location of workplace at time of onset), the questionnaire contained items designed to measure preceived job stress along a variety of dimensions (e.g., unwanted overtime, quantitative work overload, role ambiguity, lack of social support, boredom, etc.). Life stress experiences outside the job situation were measured via the Holmes-Rahe Social Readjustment Scale. Sociometric (e.g., friendship choices at work, most frequent interpersonal contacts, etc.) and personality/psychodiagnostic (e.g., Eysenck Personality Inventory, MMPI Hysteria, Hypochondriasis and Depression Scales) measures were also included. Mass illness episodes at eight distinct work settings have been investigated by NIOSH using this protocol.

The remainder of this chapter summarizes what is presently known about contagious psychogenic illness from published and unpublished reports in work settings. Table 3.1 presents summary descriptions of 15 investigations of industrial contagious psychogenic illness presently available to the authors in addition to the 8 incidents studied by NIOSH. The original reports vary considerably in terms of thoroughness, nature of the data presented, and the orientation of the investigations. The NIOSH case studies are of particular value here because a standard protocol was used which permits direct comparisons among cases.

Symptomatology

A working definition of contagious psychogenic illness might be "the collective occurrence of a set of physical symptoms and related beliefs among two or more individuals in the absence of an identifiable pathogen." Implied in this definition, but not explicitly stated, is the notion that such illness is anxiety or stress induced. Table 3.1 presents a summary of the most prevalent symptoms in each of the 23 cases. As can be seen, in the majority of incidents the symptoms were of a subjective, nonspecific nature consisting of such complaints as nausea, dizziness, headache, and weakness. In general, the symptoms were transitory, being most pronounced in the original target environment and dissipating following rest at home or medical treatment (oxygen inhalation, sedation, etc.).

The symptomatology in the 8 NIOSH cases revealed some remarkable similarities. In all cases, headache was the most frequent complaint with lightheadness, dizziness, sleepiness, and weakness being in the top 5 most frequent symptoms in 6 of 8 cases (75%). However, all symptoms in the survey were highly correlated and a factor analysis of the 25 symptoms in all 8 cases revealed a single, large general factor—that is to say, no clear pattern emerged in these cases.

Frequently, the onset of symptoms, and their subsequent contagion, was preceded by a triggering event which was presumed to be causally related to the malady. Thus, the smell of a strange odor or gas (12 of 23 cases), the use of a new solvent in a work operation (Phillips, 1974), or a bug bite (Kerchkoff & Back, 1968) have been perceived by affected individuals as pathogenic sources. In still other cases, rumored or actual epidemics involving spirit possession (Chew, 1978), dyes (Maguire, 1978), or encephalomyelitis (McEvedy & Beard,

TABLE 3.1
Summary Descriptions of Mass Psychogenic Illness in Work Settings

Source	*Setting*	*Number Affected*	*Symptoms*	*Trigger*	*Investigator(s)'s Observations*
1. Chew, 1978	Television assembly plant in Singapore	84 females in workforce of 802 (90% female)	Screaming, fainting, fear, anxiety, trance states, violent acts	No clear trigger; spirit possession reported as cause	Outbreak involved only Malay females. Hysteria spread to two neighboring plants and lasted about 2 weeks. Tranquilizers did not alleviate symptoms. Medicine-man called in to exorcise spirits but did not prevent additional outbreaks.
2. Chew, 1978	Dried cuttlefish factory in Singapore	28 females, 2 males in workforce of 350	Screaming, crying, violent acts	Visions of black man with hairy arms trying to frighten them	Outbreak involved all Malays; exorcism by medicine-man did not end outbreaks which lasted 2 weeks. Symptoms spread contagiously through workplace.
3. Chew, 1978	Cadmium-nickel battery manufacturing plant in Singapore	18 females in workforce of 500 (mostly females)	Screaming, fainting, fear, violent acts	No apparent trigger	Repeated episodes of illness occurred in 10 females, mostly young women in their teens and twenties and of Malay origin. Outbreak spread to nearby plant.
4. Chew, 1978	Camera manufacturing plant in Singapore	22 female, 1 male in workforce of 3800	Headache, screaming, fainting, crying, violent acts	No apparent trigger	Outbreak lasted one week with seven workers having recurrent illnesses. Analyses of urine were negative for morphine, codeine, and methaqualine. Outbreak spread to nearby plant.
5. Chew, 1978	Telecommunication service facility in Singapore	22 telephone operators in all female workforce of 527	Fainting, screaming, violent acts	No apparent trigger	Violent workers were physically restrained and given tranquilizers. Outbreaks continued sporadically for several months.
6. Chew, 1978	Ballbearing plant in Singapore	20 females in workforce of 800	Dizziness, nausea, fainting, watery eyes	Acidic odor	Outbreak occurred in one dept. where workers were doing overtime work. Tranquilizers given; hospital tests negative. Occurred over 4 month period.

7. Folland, 1975	Manufacturing plant in southeastern USA	85 females, 59 males in workforce of 290	Dizziness, nausea, difficulty breathing, bad taste in mouth, headaches	Strange odor	Work was perceived as boring and monotonous with high production pressure.
8. Kerckhoff & Back, 1968	Dressmaking plant in southern USA	59 females, 3 males in workforce of 965	Nausea, skin rash, fainting, headaches, weakness	Bug bite	Outbreak was viewed as an expression of frustration and unresolved stress and strain in the workplace. Speed of recovery was slower in ''depression-prone'' workers. Dissemination of belief was seen as promoting the contagion of symptoms.
9. Maguire, 1978	Ceramics factory in northern England	8 females, 2 males in largely female workforce	Skin rash on face, neck, and forearms	Raw materials used	Patch tests were negative for all but first victim who had a nonindustrial dermatological disorder. No change in raw materials used in plant for 2 years prior. Rash spread as co-workers learned of first illness.
10. Maguire, 1978	Large textile plant in northern England	8 females, 8 males	Eye and skin irritation	Dyes used for nylon fabric	Patch tests were negative except for first victim who had allergy to hair dyes after repeated home use. Rash spread as co-workers learned of first illness. Rash for co-workers only mimicked that of first victim.
11. McEvedy, 1966	Textile plant in English Midlands	22 females in all female workforce of 165	Dizziness, blurred vision, stomach complaints	No apparent trigger	Clinical tests at hospital on affected workers were negative; a check of food supply revealed no food poisoning; no toxic cause for illness discovered. Illness occurred in younger women first followed by older women.
12. McEvedy & Beard, 1973	Teaching hospital in London	102 female nurses	Malaise, headaches, nausea, dizziness, palpitations	Believed outbreak of benign myalgic encephalitis	Affecteds had more hospital admissions and more sick days from work than controls. The affecteds scored higher on the N scale of Eysenck personality test than non-affecteds.

(*continued*)

TABLE 3.1 (*Continued*)

Source	*Setting*	*Number Affected*	*Symptoms*	*Trigger*	*Investigator(s)'s Observations*
13. Philips, 1974	Manufacturing plant in southeastern USA	54 females	Headache, respiratory problems, rash	New solvent with different odor	No pathologies or toxic agents were discovered. Psychogenic factors were suspected but not assessed.
14. Shepard & Kroes, 1975	Garment manufacturing plant in southwestern USA	100 females in workforce of 340	Respiratory irritation, headaches, muscle weakness, nausea	Strange odor	Work was perceived as boring and repetitive with heavy production pressure, unwanted overtime, and poor labor/management relations. Air was contaminated by lint from sewing machines.
15. Stahl & Lebedun, 1974	University data processing center in midwestern USA	35 females in workforce of 60 females	Nausea, vomiting, dizziness, fainting	Strange odor	Work was repetitive, highly structured, and boring. Poor relations with supervisor and strict work and dress codes reported. Construction work in area produced considerable noise.
16. NIOSH- TA-76-102 (Colligan et al., 1979)	Electronics assembly plant in midwestern USA	51 females in workforce of 330	Headache, lightheadedness, dizziness, weakness, sleepiness	Strange odor	Work was perceived as repetitive and boring; high work pressure, physical discomfort, and poor relations with supervisors in affected individuals. Affecteds had higher absenteeism, witnessed others ill during outbreak more frequently, were more likely to see M.D. if feeling poorly, and scored higher on Hysteria scale compared to non-affecteds. Non-affecteds scored higher on Eysenck Extraversion scale than affecteds.
17. NIOSH HHE-77-27-437	Lawn furniture assembly plant in midwestern USA	48 females, 3 males in workforce of 356	Headache, bad taste in mouth, dizziness, lightheadedness	Strange odor	Affected workers were more often bothered by high work pressure and physical discomfort at work, had higher pre-illness absenteeism and reported more somatic complaints during an average month, and scored higher on the Hysteria scale of MMPI than nonaffecteds.

18.	NIOSH TA-77-35	Fish packing plant in midwestern USA	33 females, 2 males in workforce of 168 (day shift)	Headaches, trouble breathing, dizziness, weakness	Carbon monoxide fumes	Measured carbon monoxide levels were below toxic levels. Affecteds reported more work pressure, physical discomfort at work, poorer general health prior to outbreak, lower personal income, more dissatisfied with overtime pay, more affected by job layoffs, and perceived less staff support than non-affecteds.
19.	NIOSH TA-78-58	Shoe manufacturing plant in southeastern USA	35 females, 1 male in workforce of 400 (60% female)	Headache, weakness, lightheadedness, dry mouth, nausea	Foul odor	No evidence of toxic agent in workplace after extensive tests. Affecteds reported more unwanted overtime, physical discomfort, poor relations with management, and more financial pressure to work than non-affecteds. (Low response rate to questionnaire affects interpretation in this case).
20.	NIOSH TA-78-10	Manufacture & assembly of electronic smoke detector plant in midwestern USA	66 females in workforce of 400 (90% female)	Headaches, lightheadedness, dry mouth, dizziness, sleepiness	Foul odor	Affecteds reported more work pressure, physical discomfort, poorer pre-illness health status, more absenteeism, poorer relations with management, more boredom, and scored higher on the Hysteria and Hypochondriasis scales of MMPI. Affecteds were more likely to see M.D. if feeling poorly.
21.	NIOSH HHE-78-116-557	Garment storage warehouse in eastern USA	53 females, 19 males in workforce of 225 (80% female)	Headache, lightheadedness, dry mouth, sleepiness, dizziness	Foul odor	Small scale illness outbreaks had occurred sporadically for 8 months after moving to a new facility. Affected reported more work pressure, physical discomfort, poorer pre-illness health status, poor relations with management, higher absenteeism, and were more likely to see an M.D. if feeling poorly. Affecteds scored higher on the Hysteria and Hypochondriasis scales of the MMPI, and witnessed other workers become ill during the outbreak more than non-affecteds.

(*continued*)

TABLE 3.1 (*Continued*)

Source	*Setting*	*Number Affected*	*Symptoms*	*Trigger*	*Investigator(s)'s Observations*
22. NIOSH TA-78-8	Shoe manufacturing plant in southern USA	92 females, 29 males in workforce of 485 (60% female)	Headache, sleepiness, dry mouth, weakness, dizziness	Strange odor	Affecteds reported more work pressure, physical discomfort, poor labor/management relations, unwanted overtime, and more job dissatisfaction than nonaffecteds. Affecteds scored higher on Hypochondriasis and Impulsivity personality scales.
23. NIOSH HHE-78-138-580	Carburetor and small engine assembly plant in mid-western USA	49 females in workforce of 160 (95% female)	Headache, dry mouth, lightheadedness, sleepiness, bad taste in mouth	''Gassy'' odor	Work was manual and highly repetitious. Affecteds reported more work pressure, poorer overall health status, poor relations with management, more concern about layoffs, and scored higher on the Hysteria scale of the MMPI compared to nonaffecteds.

1973) have been offered as explanations for the exhibited symptoms. Clouding the diagnostic issue is the fact that most hypothesized causes appeared to fit well with the symptomatology and offered a plausible explanation for the observed illness. Despite negative results from medical and biochemical testings, the affected individuals in all cases displayed actual physical symptoms of illness and the related beliefs provided identifiable (and potentially manageable) explanations for their discomfort. This is a critical factor in all such incidents because most people feel a strong need to associate illness with a physical vs. a psychogenic cause.

Characteristics of the Affected Workers

Sex: The most obvious characteristic of the affected workers was their sex. Of the 1223 individuals expressing symptoms in Table 3.1, 89% were female. It should be noted, however, that this percentage is undoubtedly inflated by the fact that women were disproportionately represented in the overall population of nonaffecteds as well as affecteds. The significance of this finding is discussed in detail by other contributors to the present volume (e.g., McGrath; Singer et al.; Kerckhoff); nevertheless a few comments seem warranted here.

An examination of the work settings described in Table 3.1 reveals that the illness incidents tended to involve jobs that could be characterized as boring and repetitive, rigidly paced, with little variability or opportunity for advancement. One could argue that these are the very types of jobs typically assigned to unskilled females seeking entry-level employment. The disproportionate number of women workers reporting symptoms, therefore, might not be a function of any endogenous sex-role characteristics, but may be an artifactual correlate of a more direct relationship between job stress and well-being. This explanation doesn't negate the fact that men may also find themselves in repetitive, rigidly paced jobs (e.g., auto assembly), but one suspects that the organizational climate and parity between labor and management are much "healthier" in these settings than is usually evident in the more traditional female-intensive industries.

An alternate explanation might be that whereas male and female workers experience equivalent levels of job-related stress, working women are subject to unique "off the job" stressors which result in an overall increase in illness susceptibility. The expression of symptoms in the workplace, then, is not a response to work stress per se, but rather represents the cumulative effects of more general life stress. An important consideration here would be the extent to which the affected women evidence role strain arising from the continuing conflicts of job demands and perceived domestic responsibilities. Mandatory overtime, for example, may be perceived as a mere irritant by the working male, but is likely to be catastrophic to the working woman who feels pressured to simultaneously assume full responsibility for such domestic activities as shopping, laundering, and meal preparation.

It could also be argued that women have been socialized to express and cope with stress differently than males. A number of researchers (Jourard, 1964; Kagen, 1969; Parsons, 1955) have described the expressive vs. instrumental nature of traditional male-female roles in American society. Males have been characteristically taught to develop those behaviors most conducive to goal attainment or problem solution (e.g., ambition, assertiveness, self-reliance, aggressiveness, individualism) whereas females have been encouraged to be sympathetic, loyal, submissive, and emotionally expressive, behaviors conducive to smooth interpersonal interactions (Bem, 1974). Jourard (1964) has demonstrated consistent differences between men and women in their sensitivity to, and expression of, underlying emotions and feelings. He has shown that men typically tend to be stoic or repressive, whereas women are much more willing to externalize motivational and emotive responses. Viewing transient psychogenic illness as the overt manifestation of psychological stress or anxiety, one might expect those individuals who have been taught to subscribe to the traditional emotionally-expressive female role to be most susceptible. This point is also made by Kerchkoff and Back (1968) in their report of a case of mass psychogenic illness in a garment manufacturing plant in the south. They suggest that, in contrast to the nonaffected workers, the affecteds seemed to focus on their emotive or physiological responses (e.g., fatigue, muscular tension, headache) to the exclusion of potential external sources of stress (e.g., workpace, work overload, etc.). Thus, their attention appeared to be emotionally directed rather than problem-oriented. One suggestion for future research would be to investigate the relationship between identification with traditional sex-roles and susceptibility to, and severity of, expressed symptoms.

Personality Characteristics: With respect to the personality characteristics of the affected workers, research has been rather sketchy and inconclusive. The most commonly used psychological instrument has been the Eysenck Personality Inventory (EPI), which measures personality along the dimensions of extraversion-introversion and neuroticism-stability. The popularity of this scale is based primarily upon Eysenck's work with clinically diagnosed hysteric personalities in which he found hysterics to be more extraverted and neurotic than normals (Eysenck, 1957; 1958). Examination of Table 3.1 reveals mixed findings when the EPI was administered to affected and control samples in investigations of mass psychogenic illness. In some cases the affecteds scored higher on neuroticism than nonaffecteds (McEvedy & Beard, 1973); in some cases lower on extraversion with no difference on neuroticism (Colligan, Urtes, Wissman, Rosensteel, Anània, & Hornung, 1979); and in other cases there were no differences along either dimension (Cases #17–23).

At least part of this inconsistency may be due to the fact that contagious psychogenic illness is a distinct phenomenon, unrelated to "clinical hysteria," the "hysterical personality," or "conversion reactions." In fact, these terms have been specifically avoided in this chapter in an attempt to escape the semantic as well as theoretical confusion associated with their usage (Alarcon, 1973; Blinder, 1966; Chodoff, 1974; Cleghorn, 1969). As argued by Mayou (1975) and McEvedy and Beard (1973), contagious psychogenic illness appears to be a *social* phenomenon affecting a certain proportion of a *normal* population under conditions of psychological and/or physical stress. This does not imply that affected individuals do not differ from nonaffected individuals along certain state/trait personality dimensions (e.g., anxiety levels, coping strategies, sex-role identification, etc.) which are yet to be defined. It does seem premature, however, to expect that the characteristics of those individuals becoming ill in a normal population under conditions of considerable stress and suggestibility, as exist in a contagion situation, will parallel those of institutionalized clinically diagnosed hysterics.

One finding that does appear worth pursuing, based on the cases reported in Table 3.1, is the tendency of affecteds to have a history of absenteeism from work prior to the contagious outbreak. McEvedy and Beard (1973), describing an outbreak of psychogenic illness among nurses in a London hospital, reported that the affecteds had more prior hospital admissions and sick days from work than a group of matched controls. Investigating an incident in a small midwestern electronics plant, Colligan et al., 1979, found that affected workers had taken significantly more sick days than nonaffected workers employed in the same plant. This finding was replicated in later NIOSH investigations (#17, 20, 21).

Psychosocial Characteristics of the Contagious Environment

The work environments in which many of the illness outbreaks occurred were similar in some important respects. As a general comment, the work setting is formally structured involving well-defined roles and regimentations. Membership in these communities is neither completely voluntary nor transient. Financial need requires the individual's active participation on a relatively continuous basis. This is also true for nonindustrial cases of mass psychogenic illness such as school settings or communities (Colligan & Murphy, 1979; Sirois, 1974). Stresses or conflicts associated with such environments, therefore, may be perceived as chronic, and the avenues of escape, extremely limited.

Aside from these general organizational factors, several specific qualities of the work environment that may help precipitate such illness outbreaks are detailed below:

Boredom: A common feature of a number of the worksites listed in Table 3.1 is the low status, highly repetitive and routinized nature of the job duties that workers performed which resulted in considerable boredom and monotony (#1-7, 14-23). In the remainder of cases, such factors were not specifically reported but might be inferred (#8–11, 13) except for the McEvedy and Beard (1973) report.

Although a number of researchers have demonstrated considerable individual variability in susceptibility to boredom (Blood & Hulin, 1967; Siegel, 1969; Smith, 1955), these work conditions closely parallel the objective criteria traditionally used to define monotony and mental fatigue (McFarland, 1953; Tidwell & Sutton, 1954; Turner & Lawrence, 1965). Boredom, in turn, has been related to a variety of indices of psychological and physical well-being, including muscular tension (Cofer & Appley, 1968) overall job dissatisfaction, depression and somatization (Caplan, Cobb, French, Van Harrison, Pinneau, 1975) and increased catecholamine secretion (Frankenhaeuser, 1971).

Work Pressure: One of the more reliable correlates of affected status has been a work pressure factor. In 12 of the 23 reports listed in Table 3.1, *specific* mention is made of the fact that workers were under considerable pressure to increase production/work output. In five of these cases (#6, 14, 18, 19, 22), the strains of heavy production demands were apparently exacerbated by considerable unwanted overtime. In the NIOSH case studies (#16–23), the work pressure factor was addressed by a variety of items which dealt with worker attitudes toward workload, workpace, and time/production pressure. Sample items dealt with feeling hard pushed to get the work done, feeling unable to relax while at work, and experiencing recent sharp increases in the workload. Of the factors that appear to correlate with the experience of mass psychogenic illness the stress of work pressure warrants attention as a potential predisposing or enabling condition.

Physical Stressors: There is considerable evidence to indicate that various physical characteristics of the work environment may have contributed to the workers' overall level of stress. Although many studies do not report whether such characteristics were evaluated, the ones that do indicate their significance. For example, Stahl and Lebedun (1974) reported that keypunch operators who developed symptoms of nausea, vomiting, dizziness, and fainting at a midwestern university data processing center complained of excessive noise from dynamite blasts and pneumatic hammers at a nearby construction site. Noise was also a frequent complaint in an episode among furniture assemblers described by Colligan (Table 3.1, #17). Shepard and Kroes (1975) found discomforting levels of noise and air contamination from dust and lint generated by sewing machines in their investigation of an incident at a garment manufacturing plant.

In each of the NIOSH investigations (except the last entry in Table 3.1), physical discomfort at work correlated significantly with affected status. Physical discomforts included being bothered by poor lighting, excessive noise, temperature variations and foul odors or dust in the air. The most common complaint of affected workers in this regard was the presence of strange or foul odors (13 of 23 cases) from a variety of sources (e.g., solvents: Phillips, 1974; sealers: Folland, 1975; exhaust: Colligan, 1978; Stahl & Lebedun, 1974; pesticides: Kerchkoff & Back, 1968; Shepard & Kroes, 1975; glues: Colligan & Murphy, 1979; Murphy & Colligan, 1979). In each of these cases, workers reported being bothered by some combination of physical agents which they often considered the cause of the illnesses.

Labor-Management Relations: Poor labor-management relations were specifically noted by investigators in 7 of the 23 incidents presented in Table 3.1. The nature of the problems ranged from discontent over rigid dress codes and supervisory techniques (Stahl & Lebedun, 1974), to the lack of a clear organizational structure (Kerchkoff & Back, 1968; Shepard & Kroes, 1975). In 2 of the incidents in which the senior author has been involved (Case #16 and 17), workers had anonymously filed complaints with the Occupational Safety and Health Administration approximately one year prior to the outbreak, protesting various conditions in the work environment.

At a more basic level, there is some evidence to indicate strained or ambivalent relationships between the affected workers and their first line supervisors. Kerchkoff and Back (1968) reported that, compared to nonaffected workers, affected workers were more reluctant to discuss work-related problems with their supervisors. Shepard and Kroes (1975) placed considerable emphasis on supervisory conditions in their analysis of an incident in a garment manufacturing plant. Two years prior to the outbreak of illness, the company had replaced the female supervisors with inexperienced males in garment production. Although generally liked by their employees, the new supervisors were perceived as naive and inexperienced about sewing operations and of little help in solving or understanding work-related problems. Finally, in 5 of the 8 NIOSH investigations, poor relations with supervisors/management correlated directly with affected status. This variable included stresses from having too many bosses, receiving conflicting orders from supervisors, feeling that management did not encourage innovation or personal growth, and general dissatisfaction with supervisor(s) or management regulations.

Lack of Communication: Most of the industrial incidents described in Table 3.1 occurred in relatively noisy environments (e.g., sewing operations, keypunching). Compounding the stress due to the physical effects of noise is an impairment of interpersonal communication. This point was noted by Stahl and Lebedun (1974) who observed that communication between the various keypunch operators (and especially friends) was limited by the noise of the machines and the nature of the work. The only opportunity for communication was provided by two 15-minute coffee breaks and during lunch. Even here, workers took their coffee and lunch breaks at assigned times, often precluding sharing these times with friends. Shepard and Kroes (1975) found little social interaction among the sewing machine operators in their investigation of the garment manufacturing plant and reported that many of the affected workers expressed feelings of loneliness and isolation. Kerchkoff and Back (1968), as a result of a sociometric analysis, determined that the first five of the 62 affected workers they surveyed were social isolates, i.e., not chosen by anyone else in the sample (comprised of both affected and nonaffected workers) as friends.

In 7 of the cases reported in Table 3.1, mention was made of the spread of symptoms along social networks. Maguire (1978) noted the importance of workers witnessing or hearing about the first worker to become ill in promoting the contagion of symptoms (cases #9, 10). Likewise, Kerchkoff and Back (1968) and 3 NIOSH studies (#16, 19, 21) found evidence of a stronger social network among affecteds compared to nonaffecteds.

The role of sociometric factors in the etiology of mass psychogenic illness is unclear. On one hand, a case can be made for the spread of symptoms along friendship lines according to such traditional social psychological principles as empathy, imitation, social learning and social comparison. On the other hand, an equally good case can be made for the vulnerability of social isolates who, because of their lack of social support and belongingness, might be expected to experience greater overall anxiety and concomitant physical strain. It might be helpful to keep in mind that the *experience* of symptoms and the *expression* of symptoms are actually distinct processes which may be influenced by different sets of conditions or circumstances. Social support, for example, has been shown to be a potent moderator of both the physical (e.g. Cobb, 1974) and psychological (e.g. Caplan et al., 1975) impact of stress upon the individual. In terms of symptom development, one might expect the social isolate to be more directly affected by the physical, psychosocial, and organizational stressors of the work environment than the socially embedded individual, the latter being more reactive to the sociometric characteristics of the epidemic. As Cobb (1976) has suggested, social support is a complex construct consisting of (at least) three independent components: (1) an affective element; (2) a self-esteem element; and (3) an informational element. The affective and self-esteem element provide individuals with a feeling of cohesion or purpose and motivation to attempt to cope with experienced stressors. The informational element provides the individual with a conceptual framework for interpreting stress in terms of its probable cause and suggest specific strategies for coping with it.

Although systematic sociometric data are weak or nonexistent in many of the reports presented in Table 3.1, there is some tentative evidence to indicate that the affected workers received little informational support prior to the outbreak of illness. Thus Kerckhoff & Back (1968), in summarizing the characteristics of the affected workers, state that "affected women are more likely to have this type of (role) conflict and to deny its significance. Such denial is therefore an important correlate of becoming affected [p. 84]."

Stahl & Lebedun (1974) likewise noted a reluctance on the part of the affected workers to discuss work-related dissatisfactions. In this case, however, organizational sanctions prohibited exchange regarding job dissatisfactions. In fact, one worker was admonished for openly looking for another job.

This analysis of the social support process deserves greater scrutiny in future investigations. Based upon the available data it appears that affective or value supports, by themselves, may not be sufficient to lower susceptibility to mass psychogenic illness. In fact, by establishing greater cohesion and empathy, these factors may contribute to the contagion of illness. Informational support, on the other hand, which accurately focuses on the sources of stress and suggests effective coping strategies and alternative means of adaptation, may produce an immunizing effect.

Social support has been shown to be a potent moderator of both the physical (Cobb, 1974) and psychological (Caplan et al., 1975) impact of stress upon the individual. Given the lack of ability to communicate with fellow workers, it appears that a considerable number of workers at each site may have experienced stress and concomitant strain prior to the outbreak of observable symptoms without any social outlets. This lack of social support may have served to exacerbate the overall level of stress felt by the employees, and resulted in subsequent feelings of isolation and uncertainty among the individual employees which led to a social display of symptoms.

The Dynamics of Contagion: In their discussion of collective behavior, Milgram and Toch (1969) distinguish between "contagion" and "convergence." Contagion refers to the spread of affect or behavior from group member to member, one person serving as the stimulus for the imitative act of another. Convergence refers to situations in which group members have, independently of one another, developed common affects or response patterns that are expressed simultaneously. Although not mutually exclusive, it appears that the initial stages of mass psychogenic illness more closely parallel the process of convergence than contagion. This view is expressed by Kerchkoff and Back (1968), who suggest that the physiological and psychological arousal experiences by the affected workers is a direct response to life and job stressors. The *expression* of these symptoms, however, is facilitated by the triggering event, which offers a justification or explanation for these feelings and legitimizes their externalization. Thus the arousal, and its physical and psychological concomitants, is present beforehand and it is the "belief" in an external cause which contages.

This approach is consistent with the work of Festinger (1954), who proposed a social comparison process whereby an individual evaluates his own feelings, attitudes and beliefs by comparing them with others whom are perceived to be similar to himself. As the individual's uncertainty about his own state increases so does his need for clarification through the comparison process. Support for this position is provided by Schachter and Singer (1962) who hypothesized that an emotional response is comprised of two components: (1) a physiological/arousal response, and (2) a cognitive/evaluative response. An individual who suddenly experiences an increase in arousal in the absence of an identifiable cause experiences considerable uncertainty and looks to the environment for an explanation of what is happening to him or her. To the extent that a plausible explanation is available, an individual's feeling are labeled accordingly and an accompanying reduction in uncertainty is experienced. Thus, Schachter and Singer injected subjects with epinephrine while keeping them uniformed of its consequences (increase in heart beat, respiration, etc.). Following the injection, subjects were exposed to a confederate of the experimenter, posing as a fellow subject, who behaved in either a euphoric or angry manner. When later asked to describe their own feelings in response to the experiment, it was found that subjects tended to label their own emotional responses in accord with the cognitions provided by the model. Subjects exposed to an angry model described their own emotive responses as anger, whereas subjects exposed to a euphoric model described their feelings as euphoric.

Given the overall level of stress described in the work sites presented in Table 3.1, it is quite possible that a considerable number of workers developed psychosomatic ailments and complaints of a general nature at some time prior to the "outbreak." The expression of symptoms by the initiator, then, served to confirm the ill feelings of the remaining workers and reified their fears that something was wrong. As rumors of the illness spread through the target setting, the expression of symptoms became a legitimate activity and the number of affected workers increased accordingly. This point was emphasized by Kerchkoff and Back (1968), who observed that the first few workers to become ill evidenced the most dramatic symptoms, but as the epidemic continued and the illness gained in "respectability," workers began seeking medical attention with less severe symptoms. Even nonaffected workers may play a role in the epidemic by participating in the spread of rumors concerning the nature and cause of the illness. Thus, Colligan et al. (1979) found that although affected workers reported *witnessing* more workers become ill than nonaffecteds, nonaffecteds reported *hearing* of more people become ill than affecteds.

Summarizing what is presently known about the dynamics of contagion, one is forced to agree with Kerchkoff and Back (1968):

> The most general statement that can be made is that it (contagion) is likely to occur when a number of people are in a stressful situation for which no traditional forms of solution appear appropriate and within which develops a belief in an ambitious threatening force. There are likely also to be channels of communication among the persons involved but no organized attempt to deal with the source of strain and no general mobilization for flight. Greater specificity as to what conditions lead to this form of collective behavior does not seem possible at present [p. 3].

SUMMARY AND CONCLUSION

Although there is reason to believe that the actual incidence of mass psychogenic illness is considerably higher than previously suspected, present understanding of this phenomenon has been limited by the lack of a systematic data base. The spontaneous nature of the illness and the absence of a formal communication network have impeded the development of a standard research paradigm for the investigation of such incidents. Despite these limitations, available reports of mass psychogenic illness in industrial settings have consistently identified certain conditions as potential precipitating factors. These factors include sociodemographic characteristics (e.g., sex), nature of the work activity (e.g., highly repetitive, routinized, boring jobs), uncomfortable physical environment (e.g., noisy, temperature variations, dust/odors in the air), and psychological stressors (e.g., high work pressure, poor labor/management relations). The most powerful predictors of affected status were *not* individual/personality factors but negative reactions to the physical and social work environment. Accordingly, future research should attempt to explore these variables systematically to permit generalization across incidents. The NIOSH programmatic effort is a first attempt to utilize a standard protocol to evaluate such outbreaks and, consequently, to generate a data base or archives which would be used to detect, prevent, or ameliorate the impact of mass psychogenic illness outbreaks.

In concluding this chapter, we take the opportunity to highlight troublesome issues in this type of research:

1. Frequency of MPI—It should not be inferred that all cases of epidemic illness in industry are psychogenic phenomena. On the contrary, outbreaks due to exposure to chemical agents are by far the most common—MPI is the exception, not the rule. A "diagnosis" of mass psychogenic illness in any one case must be made prudently, only after extensive environmental tests prove negative, and based upon the integrated findings of all members of the investigative team.

2. Interpretation of MPI—To describe an illness as psychogenic does not detract from the seriousness of the illness—the illness is *real* and the workers are *sick*. The analogy to a stomach ulcer is *apropos*: The fact that an ulcer is most likely stress-induced in no way minimizes the severity of the condition; there is still a lesion in the stomach wall. Moreover, to suggest that psychogenic factors played a role in the etiology of the illness does not settle the issue. As in cases of toxic exposure, eliminating the source is the next necessary step. In the case of MPI, this may mean making the workplace more comfortable, improving the social environment at work, and/or becoming more aware of specific sources of stress/strain in the plant.

3. Nature of MPI—Mass psychogenic illness is a political as well as behavioral phenomenon—neither employees nor management appear well equipped to deal with the concept of psychogenic illness. Too often, affected individu-

als are "blamed," e.g., the illness is "all in their heads." This is paradoxical because the data suggest that MPI occurs in *normal* individuals in response to stressful organizational/work environment factors, not personality factors. Also, since MPI has now been defined, it is open to abuse. MPI sbould not be diagnosed without the extensive environmental tests that are typically employed by industrial hygienists. One form of abuse would be for management to diagnose MPI casually without sufficient environmental tests in the interests of continuing production. This situation would be more than unfortunate, it would be disastrous and extremely unfair to the workers. MPI could also be used to berate affected individuals as neurotic or "crazy." No existing data support this notion by any stretch of the imagination.

Although the potential for abuse of MPI is great and perhaps too tempting, the benefits of delineating this "health hazard" are greater yet. The research clearly demonstrates the powerful impact of job stress on individual health status. Moreover, MPI reflects one consequence of not attending to or ignoring the basic needs of the worker as an individual. Improving the quality of working life is rightly becoming a basic tenet of organizational practice and MPI serves to highlight its importance.

REFERENCES

Alarcon, R. D. Hysteria and the hysterical personality: How come one without the other? *Psychiatric Quarterly,* 1973, *41,* 258–275.

Bandura, A., & Walter, R. *Social learning and personality development.* New York: Holt, 1962.

Bem, S. L. The measurement of psychological androgyny. *Journal of Consulting and Clinical Psychology,* 1974, *42*(2), 155–162.

Blinder, M. G. The hysterical personality. *Psychiatry,* 1966, *29,* 227–235.

Blood, M. R., & Hulin, C. L. Alienation, environmental characteristics and worker responses. *Journal of Applied Psychology,* 1967, *51,* 284–290.

Caplan, R. D., Cobb, S., French, J. R. P., Van Harrison, R., & Pinneau, S. R., Jr. *Job Demands and Worker Health:* Main Effects and Occupational Differences. (HEW Publication No. (NIOSH) 75-160) Washington, D.C.: U.S. Government Printing Office, 1975).

Chew, P. K. How to handle hysterical factory workers. *Occupational Health and Safety,* 1978, *47*(6), 50–53.

Chodoff, P. The diagnosis of hysteria: An overview. *The American Journal of Psychiatry,* 1974, *131*(10), 1073–1078.

Cleghorn, R. A. Hysterical personality and conversion: Theoretical aspects. *Canadian Psychiatric Association Journal,* 1969, *14,* 553–567.

Cobb, S. Physiologic changes in men whose jobs were abolished. *Journal of Psychosomatic Research,* 1974, *18,* 254.

Cobb, S. Social support as a moderator of life stress. *Psychosomatic Medicine,* 1976, *38,* 300–314.

Cofer, C. N., & Appley, M. H. *Motivation: Theory and research.* New York: Wiley, 1968.

Colligan, M. J. *An investigation of apparent mass psychogenic illness in a furniture assembly plant.* Unpublished report prepared for the National Institute for Occupational Safety and Health, Cincinnati, Ohio, 1978.

Colligan, M. J., & Murphy, L. R. Mass psychogenic illness in organizations: An overview. *Journal of Occupational Psychology,* 1979, *52,* 77–90.

Colligan, M. J., Urtes, M. A., Wisseman, C., Rosensteel, R. E., Anania, T. L., & Hornung, R. W. An investigation of apparent mass psychogenic illness in an electronics plant. *Journal of Behavioral Medicine,* 1979, *2*(3), 297–309.

Eysenck, H. J. *The Dynamics of Anxiety and Hysteria.* New York: Praeger, 1957.

Eysenck, H. J. Hysterics and dysthymics and criterion groups in the study of introversion-extraversion: A reply. *Journal of Abnormal and Social Psychology,* 1958, *57,* 250–252.

Eysenck, H. J., & Eysenck, S. B. G. *Eysenck Personality Inventory.* San Diego, Calif., Educational and Industrial Testing Service, 1968.

Festinger, L. A theory of social comparison processes. *Human Relations,* 1954, *7,* 117–140.

Folland, D. S. Suspect toluene exposure at a boot factory. Internal report, Tennessee Dept. of Health, 1975.

Frankenhaeuser, M. Experimental approaches to the study of human behavior as related to neuroendocrine functions. In L. Levi (ed.), *The psychosocial environment and psychosomatic diseases,* Vol. 1, London: Oxford Univ. Press, 1971.

Jourard, S. *The transparent self: Self-disclosure and well-being.* New York: Van Nostrand Reinhold, 1964.

Kagen, J. *Personality development.* New York: Harcourt Brace Jovanovich, 1969.

Kerchkoff, A. C., & Back, K. W. *The June bug: A study of hysterical contagion.* New York: Appleton-Century-Crofts, 1968.

Lachman, S. J. *Psychosomatic disorders: A behavioristic interpretation.* New York: Wiley, 1972.

Maguire, A. Psychic possession among industrial workers. *Lancet,* 1978, Feb. 18, 376–378.

Mayou, R. The social setting of hysteria. *British Journal of Psychiatry,* 1975, *127,* 466–469.

McEvedy, C. P., & Beard, A. W. A controlled follow-up of cases involved in an epidemic of benign myalgic encephalomyelitis, *British Journal of Psychiatry,* 1973, *122,* 141–150.

McFarland, R. A. *Human factors in air transportation.* New York: McGraw-Hill, 1953.

Mechanic, D. The concept of illness behavior. *Journal of Chronic Diseases,* 1962, *15,* 189–194.

Milgram, S., & Toch, H. Collective behavior: Crowds and social movements. In G. Lindzey & E. Aronson (Eds.), *The handbook of social psychology* (2nd ed.) (Vol. 4), Reading, Mass.: Addison Wesley, 1969.

Mucha, T. F., & Reinhardt, R. F. Conversion reactions in student aviators. *American Journal of Psychiatry,* 1970, *127*(4), 135–139.

Murphy, L. R., & Colligan, M. J. Mass psychogenic illness in a shoe factory: A case report. *International Archives of Occupational and Environmental Health,* 1979, *44,* 133–138.

Parsons, T. Family structures and the socialization of the child. In T. Parsons & R. F. Bales (Eds.), *Family, socialization, and the interaction process.* New York: Free Press, 1955.

Phillips, P. E. Internal report prepared for the Division of Health, State of Missouri, 1974.

Robins, E., Purtell, J. J., & Cohen, M. E. Hysteria in men: A study of 38 patients so diagnosed and 194 control subjects. *New England Journal of Medicine,* 1952, *246*(18), 677–685.

Schachter, S., & Singer, J. E. Cognitive, social, and physiological determinants of emotional state. *Psychological Review,* 1962, *69,* 379–399.

Shepard, R. D., & Kroes, W. H. Report of an investigation at the James Plant. Internal report prepared for the National Institute for Occupational Safety and Health, Cincinnati, Ohio, 1975.

Siegel, L. *Industrial psychology.* Homewood, Ill.: Irwin, 1969.

Sirois, F. Epidemic hysteria. *Acta Psychiatrica Scandinavica,* 1974, *252,* 7–46.

Smith, P. C. The prediction of individual differences in susceptibility to industrial monotony. *Journal of Applied Psychology,* 1955, *39*(5), 322–329.

Stahl, S. M., & Lebedun, M. Mystery gas: An analysis of mass hysteria. *Journal of Health and Social Behavior,* 1974, *15,* 44–50.

Tidwell, J., & Sutton, J. H. *Fatigue: An introduction to a concept.* Contr. NONR-126801, San Diego College, California, 1954.

Turner, A. N., & Lawrence, P. R. *Industrial jobs and the worker.* Boston, Mass.: Harvard University Graduate School of Business Administration, 1965.

II METHODOLOGICAL AND PROBLEMATIC ISSUES IN MPI RESEARCH

Researchers of MPI do not enjoy the luxuries of a well controlled laboratory free of confounding influences with ample time to systematically manipulate variables of interest and measure consequent effects. Rather, MPI outbreaks are rapid-onset, short-lived, emotionally charged phenomena which are typically investigated in the context of a health hazard evaluation. The aim of such evaluations is to determine the seriousness of a potential threat to worker safety and health, to isolate its source(s), and to propose remedial actions with dispatch. The actual investigation is conducted *post facto*—the outbreak has already happened—and this severely limits the range of usable methodological strategies. Investigators usually are forced to rely on *current* industrial hygiene results, interviews and questionnaire responses from affected and non affected individuals, and extant company medical records. The interpretation of results from the first 2 data sets is complicated by the "after the fact" nature of the investigation and company medical records are usually of too poor quality and/or detail to pinpoint the cause(s) of the illness outbreaks.

The attitudes and reactions of employees regarding the illness outbreak also directly impact the investigation. In one NIOSH case study, a portion of the workforce decided not to enter the plant until management discovered and alleviated the causative agent and were, therefore, unavailable for individual questionnaire distribution. In other cases, individuals did not complete or return questionnaires which lowered response rate and, to that extent, impuned the representativeness of the sample. In still other cases, some individuals only answered the most general of questions (e.g., demographic items) but not the more specific work environment or personality items.

In addition to these methodological issues, there are significant practical and political concerns embodied in such investigations which serve to create adverse working conditions for the investigative team. Typically, the labor/management relations at the plant are deteriorated, due in part to the MPI episode, but to some degree by unresolved past disputes. The strained relations are sometimes evidenced by open hostility and usually by mutual distrust. The appearance of investigators on site is met with conspicuous uneasiness since all involved parties have a vested interest in the outcome of the investigation. For example, if no causative agent is uncovered and MPI is diagnosed, affected workers may be ineligible for compensation for medical expenses or lost worktime which they would have been entitled to had a specific physical agent been discovered. Plant management usually does not perceive a need to take corrective action when MPI is diagnosed and erroneously concludes that there was no problem since the illnesses were "all in the workers heads." Indeed a problem did exist and a number of people became ill, even though the cause could not be neatly defined as "agent X."

A diagnosis of MPI is always made with considerable hesitation and only when the obtained data point to powerful psychosocial factors operating in the situation. As the number of MPI cases increase, one can more confidently relate current data to this data base in seeking confirmation/justification of the MPI diagnosis. Needless to say, causative environmental and medical factors are ruled out *before* the MPI diagnosis is even considered.

The respective union and/or workforce at the plant perceives MPI as a dangerous diagnosis open for abuse by management. For example, management may forego intensive hygiene studies and diagnose MPI too rapidly in future cases, when in reality, a toxic agent may be present. In this view, MPI represents an "easy out" for management who may not see the need to initiate efforts to eliminate potential safety and/or health hazards at the workplace. Moreover, management might use the MPI diagnosis to berate workers (e.g., they're neurotic). These are viable concerns which must be dealt with through careful debriefing of all parties regarding the nature of MPI and the contribution of psychosocial as well as physical work conditions to worker safety and well-being.

The following four chapters address such methodological and problematic issues in MPI research. Dr. McGrath undertakes a careful examination of available MPI research and presents an insightful critique of both conceptual and methodological approaches currently advocated. The paper points to specific "soft spots" in this research and lists recommendations for strengthing each.

Dr. Schmitt evaluates statistical methodologies for the analysis of the eight NIOSH case studies comparing the strengths and weaknesses of various individual and aggregate analysis formats. A rationale for using the organization, not the individual worker, as the proper unit of analysis is also presented and discussed.

Dr. Boruch assesses the quality of data collected in MPI case studies and voices concern over the type of data usually collected, non-verification of self-report responses, and incomplete descriptions of methods used in published reports.

The consequences of these weaknesses for interpretation of results are enumerated and recommendations for data base management in the establishment of a national archives on MPI are set forth.

Finally, Dr. Singer presents an informal survey of troublesome issues frequently encountered in the conduct of MPI investigations. Among the topics discussed are the possible synergistic effects of physical and psychosocial factors in the etiology of MPI, political ramifications of MPI, and the correct role of psychologists in such studies. Also, the "affected worker as victim" logic in MPI outbreaks is discussed.

Taken together, these chapters represent a pleasant mix of professional viewpoints of MPI addressing scientific and practical issues and sets the stage for more theoretical treatments of this phenomenon.

4 Complexities, Cautions and Concepts in Research on Mass Psychogenic Illness

Joseph E. McGrath
University of Illinois

ABSTRACT

This chapter tries to bring some conceptual and methodological ideas from other research areas to bear upon research on mass psychogenic illness (MPI). It first deals with certain methodological and conceptual complexities involved in studying mass psychologenic illness in organizational settings. Then, it notes some cautions that may help the MPI researcher avoid several more or less generic conceptual and methodological pitfalls. The third section tries to lay out some key concepts that may help build a model of the MPI problem, and to integrate those concepts and key findings into a single coherent formulation.

This chapter is an attempt to bring some conceptual and methodological ideas from certain other research areas to bear upon research on ''Mass Psychogenic Illness'' (MPI). Three reviews of this content area have been especially useful, both as secondary sources and as sources of useful concepts and hypotheses: Markush (1973); Sirois (1974); Colligan and Murphy (1978). Much of the substantive material of this paper is drawn from those three reports.

The first section deals with some of the *complexities* of the problem, Mass Psychogenic Illness (MPI) in organizational settings. It begins with a brief description of how the problem varies as it is viewed from different perspectives; then describes some special features of the MPI problem that make it especially difficult to deal with as a research problem; and, in view of the underlying logical structure of the research questions involved in this problem area, tries to specify just what comparisons are needed in a study in order to draw useful conclusions from it.

The second section steps back from the immediate problem and considers five very general conceptual and methodological *cautions* that can help the unwary researcher in a number of problem areas within the behavioral sciences avoid potential traps. Three of these derive from the sad fact that we behavioral scientists rely very heavily on the use of verbal labels to define our key concepts. The other two are methodological problems that reflect more or less inevitable dilemmas within the research paradigms of the behavioral sciences. One has to do with measurement of key variables; the other deals with research design in general.

The third section of this chapter attempts to lay out some *concepts* that may help build a model of the focal problem. The section begins with a very brief overview of major findings and of major theoretical approaches in this problem area. These overviews rely very heavily upon the three sources noted earlier, especially Colligan and Murphy (1978). Then, two especially central constructs, exposure and susceptibility, are examined as a basis for development of a model of the problem, and are related to the main substantive findings already reviewed. The chapter ends with an attempt to integrate all of this into a single coherent formulation.

SOME COMPLEXITIES IN THE PROBLEM

The Problem as Viewed from Different Perspectives

We are dealing here with situations of the following general class: There is a particular setting—an industrial plant, a school, or the like. During a particular, bounded period of time, but not at earlier or later times, some but not all inhabitants of that setting exhibit more or less the same set of symptoms. The set may involve both physical and psychological symptoms; for example, it may include nausea, headaches, vomiting and anxiety. Both the individuals who exhibit the symptoms, and at least some of those who do not, believe that the symptoms have been caused by certain triggering events or conditions (e.g., bug bites; a noxious odor; a viral disease). Certain other relevant people, such as setting managers, physicians, researchers, on the other hand, *do not* believe the symptoms to be causally related to the alleged triggering events or conditions, or at least not to be related to them in terms of our usual instrumental-technical-scientific-medical model or paradigm of cause-effect relations.

Such patterns of behavior are referred to, in this volume, as ''mass psychogenic illness'' (MPI). Elsewhere, they have been called mass hysteria, mental epidemics, fads, panics, and the like. They represent a ''problem'' because: (1) The symptoms produce discomfort for the persons affected by them; (2) the existence of the symptoms for a number of people provides negative publicity for the setting, at least, and may indeed ''generate'' other sufferers of the same pattern; and, (3) since the symptoms are not considered to be "real" consequences of the alleged "triggering" conditions, by persons who are either powerful in the setting or experts brought into the setting, the symptoms are also not considered amenable to effective treatment by treating the alleged causal agent. Indeed, the symptoms might not be treatable in any case.

The situation poses a problem for all parties involved: The victims, the non-victims, the setting managers, and the researchers. But the nature of the problem, and certainly the conditions for solution of the problem, differ for those different parties.

For the affected individuals, the problem is one that could be eliminated if the symptoms were to cease, and if those symptoms proved to have limited negative after-effects. For the unaffected inhabitants in the setting, the problem is a problem because they might yet be affected. From their point of view, it could be eliminated if the symptoms were to stop for the "affecteds," and if "affection" were to stop spreading. For setting managers, the problem is a problem because it is costly, in terms of lost manpower and effort, and because it may have long-run negative effects on the setting (such as recruitment difficulties). The setting managers' problem could be eliminated if the symptoms were to cease, the "spreading" were to stop, and no causal agent for which they are responsible (such as toxic substances or noise) were identified as the "actual" cause. The problem is also a problem for physicians, researchers, or epidemiologists who come to the setting to study it. For them, the problem does not cease until they, collectively, can develop and verify an "explanation" for the occurrence.

Thus, the problem is not altogether the same when viewed from different perspectives; and the conditions that represent a solution to the problem for different parties in the matter are sometimes in conflict with one another. So any clear analysis of "the problem," and potential solutions to "it," must begin by specifying *whose* problem is to be dealt with—whose axe is to be ground!

Note, for example, that from the point of view of either the researcher or the setting managers, it would not be a problem, or at least it would be a quite different problem, if any one of the following things were true:

1. If only a very few setting inhabitants had "it."
2. If "it" (that is, the symptom pattern) differed from one victim to another.
3. If the explanation of the symptoms—even the same symptoms—differed from one victim to another.
4. If it was clear that baserates inside the setting were comparable to baserates outside the setting.
5. If "experts" believed the alleged causal link.

Note also that, in all of these instances, while the problem would vanish or be much altered for both researchers and setting managers, the problem would still be the same, and equally intense, for the victims.

The problem would also vanish, or be radically altered, if any of the following things were true:

6. If everyone in the setting got it (or, of course, if no one did).
7. If the same thing had happened in this setting, with regularity, in the past.
8. If the same thing was happening simultaneously in all comparable settings.

Here, again, the problem would be eliminated or altered for the setting managers and researchers; but it would remain as serious for the affected individuals.

It is probably worth asking whether or not it would be a problem if the symptoms were defined as positive—euphoria, pleasurable, spiritual—even if those symptoms led to an equal loss of personnel time on the job.

Some Special Features of the Phenomenon as a Research Problem

The phenomenon of mass psychogenic illness, viewed as a research problem, has several special features that make it especially hard to study in a systematic, scientific sense. A case of mass psychogenic illness can only be identified *post facto*. It has to already have happened—a number of persons have to have already "caught" it—before setting managers and/or researchers can know it exists. Moreover, it can only be identified *residually*. That is, it is only after setting managers have explored and eliminated a number of physical-medical alternatives—germs, toxic substances, and the like—that they reach the conclusion that "it" is a case of mass psychogenic illness. This issue is quite complex, and also quite crucial both politically and scientifically. As long as MPI must be defined residually it is, strictly speaking, logically impossible. One cannot prove the negative case. Furthermore, as various branches of basic and applied research relevant to industrial hygiene have advanced, it has become clear that humans are susceptible to ill effects from an enormous range of substances and that in some instances such susceptibility involves long-time accumulations of exposures to tiny amounts, and may even involve interactions of a complex of conditions. Such complexities make it even more difficult to decide, at a high level of probability, that a certain shared set of symptoms is *not* due, at least in part, to some biological or chemical agents.

There has been some discussion of the need for ways to establish "positive," rather than residual, diagnoses of MPI. That would certainly be desirable, for both scientific and political reasons. But like many other "needs" in the methodology of the behavioral sciences, the strength of the need is uncorrelated with the probability of the attainment. I do not think it likely, on either logical or practical grounds, that we will develop broadly acceptable "positive" bases for defining an MPI in the near future. So, MPI researchers will probably have to continue to work with a phenomenon that is defined residually—with all of the undesirable political and scientific consequences that that entails.

A third constraining feature of MPI is also definitional. The setting managers, and the MPI researchers who have been brought into the situation by them, are committed to a definition of the "cause" that is in direct conflict with what the affected persons (and other potentially affected persons in the setting) believe to be the case. The affected persons, and others in the setting, believe that the common symptoms are caused by a particular agent or event—a bug, a toxic substance, a specific disease pathogen. The setting managers are specifically and explicitly disputing this belief when they bring in a group of "mass psychogenic illness" researchers. And those researchers, *by definition,* are siding with the setting managers in that definitional dispute.

Because of these special features of the problem, the mass psychogenic illness (MPI) researcher has several strikes against him or her before the inquiry starts. Much of the information must be gathered by retrospective methods from various sets of people, many of whom have some stake in maintaining a certain definition of the problem and its cause. The affected have a stake in maintaining their definition of the causal agent; otherwise, they are to some degree responsible for their own problem: they are deluded, gullible or malingering. Researchers will need to get some of their information from those affected individuals. They will have to do so under less than ideal conditions of subject-researcher rapport.

Other inhabitants of the setting, who have not yet become affected (but might), may also have a stake in maintaining, or at least considering, the "triggering event" theory of the outbreak. Setting managers, on the other hand, have a stake in redefining the problem as a case of "mass psychogenic illness," for several reasons. First, they are likely to have some degree of responsibility for the occurrence of the triggering event (e.g., a toxic substance). Second, they certainly have a stake in getting the problem resolved so that they do not have continuing costs to bear. The researchers—I mean those investigating this "mass psychogenic illness"—certainly have a stake in maintaining that MPI definition rather than the one held by the affected individuals. Otherwise, there would be nothing for them to investigate!

Retrospective data is difficult at best. In the MPI case, it is even more difficult to get accurate information from the people in the setting. All of this makes it even more crucial that the researcher search for convergence among many different sources of information. (There will be more discussion in the second section of this paper about the need for convergence among multiple measures.)

The *post facto* and *residual* nature of the MPI problem also impose some serious constraints on potential research designs. For one thing, it is not possible to conduct an experiment, in the laboratory experiment sense of a deliberately-imposed manipulation of levels of a key independent variable. It might be possible, though, to design what is often called a "natural experiment." It certainly seems possible, at least some of the time, to take advantage of what Campbell and his associates (e.g., Campbell & Stanley, 1963/1966; Cook & Campbell, 1976) have called "quasi-experimental designs". Cook and Campbell (1976/

1979) have identified nearly two dozen designs that can offer some degree of inferential power, even when full experimental control is not possible. Of these, some of the time-series quasi-experimental designs may not be possible for most MPI cases; but others might be. Furthermore, a number of quasi-experimental designs of correlational, regression discontinuity, and nonequivalent control group forms potentially can be used in many MPI research studies. In any event, it is important that the researcher *not* use the unfeasibility of doing a "true experiment" as an excuse for not being concerned with having appropriate controls and comparisons in the study. (This, too, is discussed in the second section of the paper.)

A Formulation of the Problem in a Comparative Context

One of the striking features of research on MPI is that it does not really come to grips with the baserate question. To paraphrase the old vaudevillian's one liner: "Compared to what?" In an MPI, *some people* (but not others), in a particular work or school *setting* (but not in other settings of apparently similar kinds), during a particular *period of time* (but not at earlier times), exhibit *certain symptoms,* which either mirror those of an index case or mirror those that would be expected to follow a certain triggering event that is believed to have occurred. It is not sufficient to consider characteristics of persons who fall victim to MPI. We must also consider characteristics of those victims in comparison to characteristics of others in the situation who did not "break out" with similar symptoms. Furthermore, we must ask why the MPI occurred here, but not in other similar settings. One easy answer to that question is that no one got the MPI symptoms in other settings because the triggering event (or index case) of the MPI setting did not happen there. But few studies even consider that question, much less actually collect the data necessary to answer it.

The problem is still more complex. Why didn't all of this occur in this setting before now? Again, the argument that no triggering event occurred earlier is implicit, but usually undocumented. Moreover, we not only need to ask why these people and not others; we also ought to ask why these symptoms and not others. In other words, why is it that *this* triggering event or index case was modeled, rather than some other event?

So, we can translate the "compared to what" question into a set of questions: Why these people? Why this setting? Why now? Why these symptoms? These questions imply that MPI studies must be done not with a single control group but with a whole spectrum of control comparisons. (This requirement is also the case for many problems in the behavioral sciences, even though it is seldom noted and even less often practiced.)

Let us consider two sets of persons, one set that falls victim (V) and the other set that does not fall victim (V′) to the MPI symptoms. We can assume that there may be some differences between those two sets—whether these be personality traits, or biographical characteristics like sex and age, or past behavioral traits like absenteeism. We ought to be able to identify a comparable two sets of people in other settings in which the outbreak did not occur. We can also identify these sets of people at an earlier time. We can identify, then, eight separate "cases":

Case A Victims (V) of MPI in setting S at time T: VST
Case B Non-victims (V′) in S at T: V′ST
Case C Those same victims (V) at an earlier time (T′) in S: VST′
Case D Those nonvictims (V′) at T′ in S: V′ST′
Case E People like the victims (V) who are in situations like S but in which MPI did not occur (S′) at T: VS′T
Case F People like the non-victims (V′) in S′ at T: V′S′T
Case G The victim-like people (V) in S′ at T′: VS′T′
Case H The non-victim like people (V′) in S′ at T′: V′S′T′

The problem for the MPI researcher is not just to describe differences between V and V′ in S, T (that is, differences between case A and case B). The MPI researcher must also ask how V and V′ differed in ST′, in S′T, and in S′T′, even though neither set showed the MPI symptoms is any of those cases. The MPI researcher also needs to account for why V in S, T (Case A) showed MPI symptoms while V in S, T′ (Case C), in S′T (Case E), and in S′T′ (Case G) did not.

The main point here is that the MPI problem implies a very complex set of comparisons. These comparisons involve individual differences, situation differences, and temporal differences. Furthermore, it is unlikely that we will solve the puzzle in terms of any simple personality difference between those who are and those who aren't affected by an MPI (that is, $V \neq V'$ on property X_1), or even in terms of differences on a whole set of individual properties (that is, $V \neq V'$ on properties $X_1, X_2, X_3 \ldots X_n$.). It is much more likely that we will come to understand MPI in terms of complex interactions of personal differences, differences in time, and differences in setting. We will explore some of the implications of these complex comparisons in the third section of this paper, after first considering some more general conceptual and methodological difficulties.

SOME CONCEPTUAL AND METHODOLOGICAL CAUTIONS

Although the MPI problem has certain special features, some of which have been noted and discussed in the preceeding section, it also has many aspects in common with a number of other areas of social psychological study. These include areas such as the study of factors in human stress, the study of the spread of innovations, and the study of the occurrence of accidents. Furthermore, the MPI area shares certain conceptual and methodological problems with all of the social and behavioral sciences. Some of these represent serious and subtle pitfalls that are particularly likely to ensnare the MPI researcher. This section is an attempt to offer certain cautions which, if heeded, could help the MPI researcher keep out of at least some of these pitfalls.

The Tyranny of Words

Several of the pitfalls, and the associated cautions, arise because much of social and behavioral science is heavily dependent on concepts formulated in words. When we use words to label constructs in social science we can mislead (ourselves and others) in essentially three ways, and those three ways are the bases of the three cautions to be listed here.

Caution # 1: Try Not to Let the Terms You Use Prejudge the Answer to the Question You Are Studying

Sometimes researchers label a phenomenon that they intend to study with words that are heavily value-laden. This can have the effect of shaping the researchers results before the inquiry starts. The study of stress provides a classic example of this problem. Researchers who study "stress" find that high levels of "it" lead to "deterioration" of performance. They seek ways to cope with "it." They sometimes try to find factors that give people stress resistance. Some other researchers, who study what seem like the same or related phenomena but do so under labels such as "activation," "arousal," or "motivation," find that increases of "it" lead to improved performance. They seek ways to help people *not* lose "it." Coping, a good thing, becomes habituation, a bad thing. They certainly don't try to find ways to make people resistant to such motivation.

Examples abound in other areas as well. Those familiar with the literature on social influence will recognize that when "conformity" vs. "deviation," as studied by Sherif in the thirties, became "yielding" vs. "independence" as studied by Asch in the late forties, essentially the same responses to social pressure changed places in terms of which was "good" and which was "bad." Social psychologists aren't the only sinners. The same reversal of good and bad meanings for a construct occurred when the religious team, resistence to temptation, was replaced by the psychoanalytic term, repression.

The MPI researcher should consider this idea very carefully. Calling these things "hysteria" has already been recognized as pejorative—and perhaps sexist. Calling such things panics, or mental epidemics or even mass psychogenic illness, also connotes some answers to the very questions being studied: What are the causes of the symptoms? How do they spread? Who is responsible?

Even though there is already a plethora of names for the phenomenon—mass hysteria, mental epidemics, mass psychogenic illness, and the like—I want to add one more. Let me suggest that we label the phenomenon *Multiple Occur-*

rences of Unexplained Symptoms (or MOUS). I am trying to choose a label that is descriptive, but that avoids prejudging the nature and causes of the phenomenon. I avoid "mass" because some of the cases covered in the reviews involved only 3 or 4 people; many involved less than 10. I avoid "contagious" and "epidemic" because they imply the answer to the question: "How do the symptoms spread?" I avoid "psychogenic" because I don't know how one decides *which* vomiting, *which* headaches, and *which* nausea are "physical," and therefore "real," and which are in some sense "unreal." I also avoid the term "illness," because some of the incidents covered in past reviews could scarcely be called that, except in metaphorical terms. The "dancing mania" of centuries past and the "rock concert" and "disco mania" of current times may differ more in our cultural interpretations of them than in manifest behaviors of the "affected." Why are the former "illnesses" and the latter "fads" or "recreation"? I even avoided the term "outbreak," which seems to me to imply that the phenomenon is entirely *nonvolitional*. I don't think even that should be decided in advance.

So, I offer the label: *M*ultiple *O*ccurrences of *U*nexplained *S*ymptoms. (MOUS!) I'd welcome a good E word at the end, to round the acronym!

In the present context, we are more interested in MOUS that (1) occur in industrial/organizational settings; (2) are interpreted by the inhabitants as involving *negative* symptoms; and (3) are interpreted by setting managers, and/or outside experts, as not having a specific pathogen and, more precisely, as *not* having been caused by what the affected (and perhaps the uneffected in the setting) believe to have been the cause. But even within these limits, what we call the phenomenon is important because that label guides our search for causes, consequences and understanding.

Caution #2: Try Not to Let Phenotypic Similarities, Especially Those Based Solely on Verbal Labels, Mask Genotypic Differences

In the social and behavioral sciences, constructs are defined in verbal terms. As the Humpty Dumpty character in *Alice in Wonderland* has pointed out, we can indeed make words mean whatever we want them to mean. But when a researcher applies a given verbal term to a given phenomenon, operation, variable or event, that researcher and others are likely to proceed on the assumption that "it" is somehow the same as everything else that has received that same verbal label.

An example from the stress area is again appropriate. The array of events and conditions that have been called stressors or stress conditions is remarkably broad and diverse. It includes:

1. night infantry combat;
2. wintering over in the arctic;

3. failing a test;
4. losing a loved one;
5. being told that other college sophomores in an ad hoc group don't like you;
6. being told that other college sophomores in an ad hoc group disagree with you;
7. undergoing potentially fatal surgery;
8. being divorced;
9. being married;
10. being out of the company of other people (in one case, an infant being out of sight of mother for a whole minute).

It is powerful magic, indeed, that enables the researcher to make all of these things somehow the same merely by giving all of them the same verbal label (and waving a Ph.D. at them as a magic wand). This problem, too, is not limited to the stress area. Consider, for example, the array of things that have been taken to be "group productivity," or "aggression," or "liking," or "attitudes," in the literature of experimental social psychology.

The MPI researcher should keep this caution in mind, too, when he or she sets out to identify, classify, or catalogue events that will be considered members of the class of phenomena being studied. Is the reaction to the Orsen Wells version of H. G. Wells *War of the Worlds* the same kind of thing as the early Children's Crusades? Are the activities of the Flagellants of European "dark ages" to be assigned to the same class of phenomena as the various polio scares of the past half century? Are the various "dancing manias" of the 10th through 16th centuries in middle Europe the same kind of thing as the modern industrial MPI? A review of the literature of this area might lead one to conclude that the answer is yes to all of these questions. At some point, a verbal label that is used to cover such a wide array of events loses its usefulness as a scientific construct. Internal differentiations may be needed. Sirois (1974), for example, describes five different types of MPI, with differentiations being made primarily in terms of the time course of the "epidemic." Perhaps there is a need for more analytic effort toward developing taxonomies of types of MPI.

Caution #3: Try Not to Let Phenotypic Differences, Especially Those Based on Verbal Labels, Mask Genotypic Similarities

This is the obverse of Caution #2. Just as merely attaching a particular verbal label to two sets of events does not make them the same thing, so merely associating different verbal labels with two sets of events does not guarantee that they are dissimilar. Social scientists have the bad habit of inventing new labels for what persons with less at stake might see as the "same old" concepts. I have always been surprised, for example, that the concept "obedience" was treated as if it were a totally new idea, by researchers in a field that had been studying conformity, deviation, authoritarianism, and other aspects of social influence for several decades. I have also been amazed at how many different terms there are for positive interpersonal regard (terms such as liking, positive affect, positive interpersonal perceptions, positive sociometric choices, cohesiveness, and the like). A colleague and I once labeled this kind of labeling as a search by the social scientist for "marginal differentiation of product" (McGrath & Altman, 1966).

Sometimes it can serve a very useful purpose for a researcher to consider the similarities between two or more sets of phenomena that have been given different names and treated as different "things." One example of such gain, from the stress literature, is in the very fruitful sets of ideas that have come about because some stress researchers decided to consider certain stimulus "under loads" (e.g., social isolation) as potentially related to task or stimulus "overload"—related, that is, in their stressing effects on the "loadee."

In this context, then, it probably is worth asking the MPI researcher whether the dancing manias of the earlier times ought to be considered as related to the current disco dancing popularity, or the recent popularity of rock concerts. Similarly, it ought to be worth asking about the similarities, if any, between cigarette smoking and heroin addictions on the one hand (both of which are considered by Markush (1973) as mental epidemics); and on the other hand, such phenomena as: (1) the relatively rapid rise in use of sugar in certain western countries; (2) the vast increase in caffeine addiction; (3) the rapid spread in television watching; and alas, (4) even our "addiction" to various forms of higher education? I am not proposing that any or all of these matters ought to be labeled as "problems," much less as MPI problems. I am suggesting that analysis of the features that these matters do and do not have in common with MPI may be well worth the effort in terms of new insights gained.

In summary, then, MPI researchers, like researchers in other social science areas, must understand that they cannot make things good or bad, similar or different, by merely labeling them. But they also must understand that, by labeling concepts in particular ways, they may mislead others, and even themselves, regarding these matters.

The Tyranny of Logic

There are several pitfalls—indeed, dilemmas—that arise from the central fact that social science is committed to the always-fallible empirical measurement of constructs, and to the always-limited and always-uncertain inductive logic of inference. Although the consequences of these pitfalls are legion, I will concentrate here on two cautions. One deals with empirical measurement and its meaning. The other deals with research design and its consequences.

Caution #4: Because All Measures Are Fallible, Try Not to Let a Concept Get Tied to One and Only One Method of Measurement

There is not enough space in this chapter to lay out in detail the ways in which measures of concepts within the social and behavioral sciences are error-laden, in both reliability and validity senses. Nor is there enough space to argue the case for multiple operationalism, which is a limited, but as yet best available, strategy for gaining confidence in both the precision and the meaning of measures. The reader is referred to the body of literature on reliability and validity of measures, and on internal, external and construct validity of empirical relationships. (See, for example, Campbell & Fiske, 1959; Campbell & Stanley, 1963/1963; Cook & Campbell, 1976/1979; Runkel & McGrath, 1972; & Webb, Campbell, Schwartz & Sechrest, 1966.)

Let me summarize one major thrust of much of that literature. Both reliability and construct validity hinge on the twin ideas of: (1) the *convergence* of data obtained from different methods of measurement that are intended to be measuring the same thing; plus (2) the *divergence* or differentiation of data from two measures that are intended to be measuring different things. This, in turn, implies that it is *necessary* to find or develop more than one method for measuring every key construct; and that such multiple methods must be as different as possible from one another in form, so that they will not share the same weaknesses or fallibilities but rather will offset each other's weaknesses.

Examples of the problem of single-method concepts are ubiquitous in the literature of social science. For example, certain personality characteristics are defined in terms of a score on a certain inventory or scale, and no other operation is considered to be a measure of the same thing. Androgeny *is* the score on the androgeny scale. Authoritarianism *is* the score on the California F-scale. Machivellianism *is* the score on the Mach scale. The multiple-operationalism advocated here is certainly honored more often in the breach than in the observance. Moreover, there are at least some reasons to suspect that a very large number of our favorite concepts in the social sciences would probably not survive the test of convergence of multiple measures if they were so tested. But no matter how infrequently the multiple-operation strategy is followed in practice, and no matter how infrequently it yields research outcomes that please us, it remains the case that convergence of evidence from different sources is *necessary* within the empirical/inductive strategy of the behavioral sciences.

For the MPI area, measurement of the dependent variable—expression of the MPI symptoms—poses special problems. Nearly always, that measurement must be done retrospectively. Use of ''subjective'' methods (self-report questionnaires, ratings or interview responses) pose all sorts of problems in any event, especially problems of reactivity. (See Webb et al., 1966, for a sophisticated treatment of these problems). These problems are aggravated a lot when self-report measures are used retrospectively; and they are literally multiplied when such measures are used under conditions where the respondant is likely to be hostile to the research. (See the first section of this paper for a discussion of the poor rapport conditions facing the MPI researcher.) These problems apply whether the researcher then uses the self-report of 1, 3, or 37 "symptoms" as the definition of "having" the MPI.

On the other hand, to simply substitute some "objective" indices of MPI, such as records of who received medical treatment for the symptoms on specific days, does not altogether solve the problems. Medical records are far from error-free, in a clerical-error sense. Moreover, the recording of symptoms is very likely to be influenced, from one case to the next, by the pattern of symptoms that seems to be "going around," so to speak. On top of that, just who does and does not actually get medical attention, under the confused, problematic, ambiguous, and even angry circumstances that often characterize MPI *in situ,* may be more or less random. Solving the reactive problems of self-reports by substituting an array of data-quality problems is not likely to be an effective tactic. Instead, the MPI researcher must learn to use *all* the measures he or she can get, and to use them together, to transcend the limits of any one of them.

The use of multiple operations to show convergence of and discrimination between constructs is one approach that can help deal with the problems mentioned in Cautions #2 and #3. One answer to the question of whether things given the same label are in fact the same is whether or not measures of them converge, or covary. One way of finding out whether two things that are known by different labels are in fact the same is by exploring whether measures of them converge or discriminate between them.

Caution #5: Do Not Let the Unfeasibility of Doing Well-Controlled Laboratory Experiments in the MPI Area Serve as an Excuse for Ignoring the Logic of Experimental Design in the Broader Sense, or as a License to Do Sloppy Research

I'm afraid that this may be one of the most prevalent pitfalls in the social sciences. Researchers who are working in what are thought of as relevant or applied problem areas often construct a strong case for why they cannot do "tight" experiments, in field or laboratory, in these problem areas. Far too often those researchers go on to conclude—or at least to act as if they had concluded—that therefore they and their studies are somehow exempt from concerns about internal validity, external validity, and the logic of inference. They seem to argue as follows: "One can't do really good research in this problem area, since experimentation is ruled out. On the other hand, this is a truly important problem area. Therefore, since poor research is better than none at all, we will just do the best we can, and not worry too much about such *exotica* of research as the logic of inference, internal validity, statistical significance and all that jazz."

That line of argument is fraught with flaws. Let me point out just a few of them. One is that poor research is *not* better than none at all. Indeed, it is infinitely worse! Someone might believe it! That is a bad enough condition for basic research areas, because it often takes many years to expunge false positives from the literature. But publishing results of poor research is much more serious, even dangerous, in applied areas, because someone might haul off and act on those results!

Second, such matters as construct validity, internal validity, precision and the like, are anything but exotic luxuries for the researcher. They are the guts of the research enterprise. Third, this line of argument seems to postulate two classes of studies: one class of more or less perfect research—experimentation—*vs.* all other, equally imperfect, research. Nonsense!! Experimentation is, by and large, filled with problems; certainly even the best of it is far from perfect research. On the other hand, there are many other forms of research, many shades and variations. Each of them offers *some* advantages, though all of them have weaknesses. The trick is not to try to plan and carry out the perfect study—it is not possible. Rather, the trick is to do the best we can in each study—while worrying a great deal about validity, precision, and the like; and then to use multiple studies in the same way that we can use multiple measures—to seek convergence across studies on our key constructs and relationships.

Within that broad prescription, and in the context of nonlaboratory research on problems such as MPI, it is vital that the researcher understand that much of the strength of any particular study hinges on just what comparisons can be made within it. All knowledge is knowledge of differences. When we purport to have established that a certain phenomenon or relationship holds for a certain body of data, we must make that assertion against the implicit comparative question: Compared to what? This idea includes the idea of baserates and baseline comparisons. It also includes the idea of control groups. Within a research problem of any degree of scope or complexity the idea of "a" control group must be greatly expanded to the more complicated ideas of a range of control and comparision conditions. These ideas, in turn, can be expanded to connect to the much more general ideas of quasi-experimental designs as put forth by Cook and Campbell (1976/1979).

So even if the MPI researcher is not able to run a true experiment, laboratory or otherwise, he or she is still beholden to develop a study plan that takes into account the need for comparison, usually a series of comparisons rather than a single comparison or control group. If the researcher does not do so, then results of the study will really be of very little use. Results will not be able to deal with the question: Compared to what?

The ideas on quasi-experimental designs put forth by Cook and Campbell offer nearly two dozen potential designs. None of them has the logical power of a true experiment, but each of them offers sufficient logical power to allow the researcher to gain some knowledge with some confidence. Not all of these will work for any given problem, especially a problem like MPI which has such special, constraining features. But *some* of these quasi-experimental designs are likely to be useful for any given problem, even the difficult and constrained MPI problems. It behooves the MPI researcher to pursue these possibilities. In a highly difficult area such as MPI research, we need all the help we can get!

SOME CONCEPTS FOR A MODEL OF MPI

The preceeding section dealt with relatively broad and abstract conceptual and methodological issues that seem critical to research on MPI. This section returns to more concrete matters. It starts with a brief summary of some of the main consistent research findings about MPI. That summary is followed by a brief discussion of some of the major theoretical approaches used in this area. Both of these rely very heavily upon material from the three references noted at the outset, especially the Colligan and Murphy (1978) review. The discussion of theoretical approaches makes it clear that there are at least two underlying constructs that are implicated in almost all of them: differential exposure and differential susceptibility. The third part of this section is an analysis of these two concepts. The fourth part of this section is an attempt to integrate these two concepts with the main substantive findings. This section and this chapter conclude with an attempt to integrate all of this material into a coherent formulation.

Sex, Stress and Somatization: A Summary of Some Consistent Findings

Sex Differences

The most consistent single finding in the literature of MPI is that females "catch" such "diseases" far more frequently than males. On the other hand, it is clear that the kinds of settings in which these outbreaks occur tend to have almost all females at the "assembly line" level of workers. It is not entirely clear whether the very high percentages of affecteds who are female (perhaps 80% to 90% or more) is really higher than the baserate of females in those or comparable settings. So the overall frequency of female victims of MPI can be interpreted as meaning either: (1) that more females than males are victims of MPI's; or (2) that MPI outbreaks tend to occur in those settings where most of the workers (or students) are females. Both formulations leave some unanswered baserate questions.

For the most part, researchers who have noticed and discussed this high incidence of females in MPI have tended to assume the first of the two statements above, and have tried to account for why females are victims of MPI more often than males. In fact, the very term "hysteria" is a label for the disorder that predicts, and at the same time "explains," the disorder as one that only females will suffer. It is said that Hippocrates used the word hysteria because he thought it was caused by the uterus, literally, wandering in search of semen. Most current researchers in this area would reject such an explanation, and indeed would probably reject the use of such a sexist term as hysteria. Perhaps part of the reason why the finding that most victims are female is given remarkably little discussion in the MPI literature is because many researchers, in this still male-dominated area of behavioral science, want to avoid seeming to be sexist.

The attempts to "account" for the high female frequency are most often built upon notions of learned differences in sex roles. At least two such explanations are offered by Colligan and Murphy (1978). One starts with the idea that females in the work setting are in conflict. They have major role responsibilities in the home in relation to family and household. They are uncertain of their role in the work place. Thus, they are under role stress, and the MPI is an expression of that stress. But if this were so, then we would expect to find that single women, and/or women with more feminist views, would be less susceptible to MPI. We should also expect to find that men who have such dual role responsibilities (male heads of single parent families) would therefore be more susceptible to MPI than other men. Such comparisons have apparently not been made; and Colligan and Murphy (1978), appropriately, call for research on this question.

An alternative explanation is based on differences in the content of traditional sex roles. The traditional female sex role, into which most adult females have been socialized, is one that emphasizes expressive (as opposed to instrumental) behavior. In this explanation, the MPI is to be viewed as an expression of (that is, a somatization of) job stress. Males, because of their sex role training, are more apt to externalize (that is, blame the work-conditions, the boss, etc.). This explanation has several problems. First, although there is considerable evidence that the victims of MPI are more likely to express stress in the form of bodily symptoms (that is, somatization; see below), it is not clear how that is related to sex role training. In most usages of the dichotomy "expressive" vs. "instrumental," the term expressive *does not* particularly refer to expressing things in the form of bodily symptoms. Nor does the term instrumental connote *projection* of stress or symptoms onto environmental targets. Moreover, if this differential sex-role explanation were sound, then it ought to be possible to find a subset of females and males who are not socialized into those alleged traditional roles (e.g., females or males who have high scores on an androgyny scale). Such a test does not seem to have been made.

The second interpretation—that MPIs tend to happen in situations where most of the workers (or students) are female—deserves a comment. Some other factors related to incidence are pertinant here. MPI seem to happen to the relatively young, to persons in relatively menial jobs, to persons in relatively unpleasant settings (see later material on job and environmental stress), and in settings where communication to peers and to superordinates is limited or difficult. Perhaps the high incidence of females, as both affected and nonaffected members of the settings in which MPI occur, simply reflects our tendency to put females in low status and otherwise unrewarding jobs. If so, then "female," along with "young" and "poor," should be treated as a variable that signals "low social status," not some character flaw like "hysteria" or "hypochondria." Perhaps the MPI symptoms represent a "last resort" means for low-status persons (female, young, poor, etc.), in high stress situations, to express their distress.

Personality

Considerable effort has been devoted to trying to find "the" personality traits that differentiate victims of MPI from those in the same situations who do not fall victim. As has so often been the case with efforts to find personality traits that predict behavior, in many other problem areas, results here have been weak, contradictory and not very helpful. The two most frequent candidates for such differential personality traits (apparently for reasons of convenience, rather than for strong conceptual reasons) have been neuroticism and extroversion. Colligan and Murphy (1978) summarize results of many of these efforts by noting that some studies show MPI victims to be high on both, some show them to be high on one but not on the other, some studies find no differences on either of them. In other words, the findings are neither consistent nor replicable.

There are some sound logical reasons why we should not expect such "traits" of the individual to be very strong correlates of displaying or not displaying MPI. Those reasons involve the obvious interactions of "who" with "where" and "when" in these MPI outbreaks. So, for both conceptual and empirical reasons, we ought to expect little help in understanding MPI from the personality-trait approach.

Illness Related Behaviors: Somatization

One relatively consistent set of findings that bear on the question of why particular individuals, and not others, fall victim to MPI, indicates that the "affected" also show higher rates of absenteeism, more sick leave days, and more medical consultations. That constellation of indicators could be signaling either of three things about victims of MPI. First, it could mean that they are more sickly, in some general sense, and hence are more likely to fall victim to the MPI (or any other) symptoms. Second, it could mean that they are more hypochondriacal, in general, and hence are more likely to imagine that they have the MPI symptoms. Third, it could mean that they are not more sickly or hypochondriacal than others, in general, but that *given high stress* they are more apt to use somatization—expressing stresses in the form of bodily symptoms—as a preferred coping style. The small amount of evidence available seems to refute the first and support the third (and perhaps the second) of these explanations (Colligan & Murphy, 1978). It seems reasonable to hypothesize—though tentatively, to be sure—that the occurrence of MPI is the somatic expression of stresses in the situation, and that those who become MPI victims are those who use somatic expression as a preferred mode for coping with such stresses. To refer back to an earlier section, it is somatization that sorts out victims vs. nonvictims of MPI, whether or not that somatization is connected to female sex role socialization or sex role conflict.

Situational Factors: Stress

Besides the efforts to find bases for predicting who will and who will not fall victim to an MPI in a situation where it does occur, some effort has been given to trying to find out why MPI occur where and when they do (that is, why in this situation, and why at this time). For one thing, there is almost always a triggering event, or an index case. The triggering event could be the company's change in production process, or in work schedules, or in the solvent it uses, or the like. An index case is the occurrence of a case of "actual" illness—that is, one that medical personnel diagnose as caused by a germ, a toxic agent or some other denotable pathogen. The MPI consists of later "cases" that (1) mirror the index case in symptoms, or offer symptoms that are reasonably compatible with effects that the triggering event might cause; and (2) are attributed by victims to the triggering event or to the same pathogen that caused the index case; but for which (3) the medical personnel involved do *not* accept that diagnostic attribution.

It also turns out that, in most of the MPI occurring in industrial settings in recent years, there is evidence of considerable negative job and environmental conditions. There are often reports of excessive noise, noxious odors, poor lighting, variations in temperature. The settings frequently involve extremely boring, repetitive tasks. In a high proportion of cases, there is evidence of poor labor-management relations: worker-supervisor friction, ambiguous supervision, heavy production pressures, and poor conditions for worker-to-worker communications.

Such results suggest a situational-stress interpretation of the phenomenon. Perhaps we can regard the relatively bad conditions of environment (noise, odors, etc.), of job (boring, repetitive, work production pressures) and of interpersonal relations (supervisor friction, poor worker communication, etc.) as representing a high level of stress in the situation. If we act as if these are substitutable one for the other, then we don't have to show that any one of them is present in all cases, only that "a bunch" of them are operating in each case. But presumably such conditions are present and having an effect on *all* workers in the situation, but only some of them become victims of the MPI. There is also the problem that such conditions are probably operating, at comparably high levels, in lots of other industrial settings, but most of those settings do not spawn MPI. Finally, there is the problem that *this* setting has probably had such bad conditions operating for quite a long time, and MPI have not occurred before. So stress appears to be a necessary part of the explanation of MPI, but not by any means the whole explanation.

Summary

These considerations point to some key questions and comparisons, already noted in the first section. We need to be able to specify some of the features that distinguish this situation and time from others like it. We need to understand the role of triggering events or index cases in ''unleashing'' certain responses, or coping styles. We need to identify characteristics on which the population of victims are *different from* nonvictims—and different from them in ways that are related to the relevant situational conditions and triggering event(s). In other words, the crucial features of MPI may be first or second order interactions, of person, situation, and temporal circumstances.

The major consistent findings seem to support three constructs, as suggested in the title of this part: sex (actually gender differences); stress (in terms of job, environmental and interpersonal conditions); and somatization (that is, the coping style preference for reacting to stress in terms of bodily symptoms). These represent a reasonable beginning, but by no means a complete set of substantive concepts. Nor do they answer some crucial theoretical questions. to which we now turn.

Some Theoretical Approaches to MPI

Some of the ''theoretical'' discussions of MPI place heavy reliance on psychoanalytic concepts. Others rely more on epidemiological and social psychological ideas. The latter include convergence, contagion, and emergent norms theories. Colligan and Murphy (1978) offer a useful discussion of some of these, and the following material relies upon that discussion. (See, also, Kerckhoff, Bach & Miller, 1965; Milgram & Toch, 1969.)

Psychoanalytic Approaches

These interpret the manifestation of MPI symptoms as reflecting some deeper, unconscious, psychological problems of those individuals. From such a point of view, the particular symptoms displayed are relatively unimportant. If they didn't exhibit these symptoms, they would exhibit some others instead. To setting managers, on the other hand, and to the victims, the symptoms *are* the problem, and getting rid of the symptoms is the goal.

Basically, the psychoanalytic approaches are neither verifiable as theory nor useful as guides to interventions. They hinge on interpretation of *all* responses as manifestations of inner conflicts. Because such underlying conflicts are only known when manifested as ''symptoms,'' and because *all* behaviors are symptoms of some underlying problems, the psychoanalytic theorist, *in principle,* has no conceptual tools to anticipate *which* persons will use MPI symptoms and which will use other "symptoms" to express their (underlying and unknowable) problems. Nor does the psychoanalytic theorist really care which sets of symptoms the individual uses (i.e., MPI or other). Rather, that theorist is concerned with ferreting out the underlying problems—the genotypic problems, for which MPI and other symptoms are phenotypic expressions. So, the psychoanalytic theorist has neither interest in nor tools to solve the MPI problem.

Convergence

The central idea of convergence is that a bunch of individuals all are exposed to the *same* index event, and are all affected by it independently of one another. The people who are affected are somehow different than others, who either were not exposed to the same sources, or were not affected by that exposure. Temporal differences in exposure are more or less unimportant variations in the relation between being exposed and being affected.

One of the psychoanalytic approaches can be viewed as a convergence theory. In it, the victims *identify with* the index case, and thus succumb to the same illness; whereas nonvictims do not identify with that particular person. Evidence supporting the idea that the index case is somehow notable, and therefore to be identified with by a number of people, is rather weak.

Contagion

The basic idea of contagion is that A, the index case, somehow transmits to B, who in turn transmits to C (and, of course, perhaps to more than one other person) and that this occurs in sequence. In this approach, the time effects are not trivial, but reflect order-of-exposure. They thus tell us a lot about the process of contagion, and about how the "disease" sweeps through a group. In the contagion approach, who gets the symptoms is not a function of their relation to the index case (except for the first one), but rather is a function of their relation to one another in both physical and social senses. We would expect job and social isolates to be least affected, task and social "stars" to be most affected.

The evidence supporting this is again weak, especially if we insist that any one MPI must be *either* contagion *or* convergence. Actually, Colligan and Murphy (1978) make a good case for the idea that convergence may fit the early part, contagion the latter part of an MPI. Kerckhoff et al. (1965) and Milgram and Toch (1969) also support that idea. At first, it is people who "identify with" or "imitate" the index case, but who are not especially related to one another, who "get" the MPI. Then, "it" gets into and spreads through social networks, in a contagion manner.

Both of these raise the question: Why does the MPI not sweep through the entire population? Perhaps one answer is that some people are in social networks that never do get "infected." Even though for some people the spread of the disease is facilitated by their ties in social networks, for others a particular network acts as a defense against the MPI.

Emergent Norm

The basic idea here is that an emergent norm develops in the group, such that getting and expressing the symptoms becomes an acceptable way of dealing with the considerable stress in the situation. (See Turner, 1964; Turner & Killian, 1957.) This can be seen as a restatement, in social psychological terms, of the more epidemiological convergence and contagion theories. Convergence could occur in an early stage before the norm exists; with contagion occurring as the basis for spread of the norm after the norm gains strength. The epidemic ceases short of a 100% victim rate because some social networks may never accept the emergent norm, thus serving as a defense against contagion.

Summary

The convergence, contagion and emergent norm approaches probably should be taken as complementary rather than as competing theories. Together they reflect a "flow" model, with different conditions affecting rate, direction and cessation of flow. There are at least two concepts that seems to be at the core of all of these approaches: the idea of differential exposure of individuals; and the idea of differential susceptibility given exposure. The next part of this section tries to analyze these two concepts as a step toward a model of the MPI problem.

Two Core Concepts for a Model of MPI

Both the main consistent findings (sex, stress, and somatization) and the main contending theoretical views (convergence, contagion, and emergent norms) all suggest that any viable model of the MPI problem will need to include at least two underlying concepts: *differential exposure* and *differential susceptibility*. The concept of exposure is basic. The whole idea of an epidemic, or contagion, implies that something is transmitted to a victim by contact. In the general case, either a toxic substance is transmitted from an environmental source to a victim, or a virus or bacterium is transmitted from one victim to another.

In the case of an MPI, it is not a pathogenic agent, but rather *information*, that is transmitted. It is a relatively complex set of information, too. An MPI victim must have been "exposed" to:

1. the fact of the occurrence of the index case, or the triggering event;
2. the fact that others have fallen victim to the illness; and
3. the symptoms involved in the illness.

The idea that the MPI case involves exposure to *information*, has several implications for MPI research. First, exposure implies contact, but not necessarily physical contact with either a person or an environmental condition. Furthermore, ''contact'' and transmission of information can occur one-to-many as readily as one-to-one. Therefore, the meaning of time intervals and sequences in the spread of ''contagion'' become less clear cut. The distinction between *convergence* and *contagion*, which partly involves time, becomes blurred. In addition, because the ''ingredient'' transmitted by exposure is a complex set of information, the exposure in MPI must be regarded as a fairly complex set of *conscious* cognitive events.

The concept of exposure in a ''contagious'' situation implies that exposure and its effects are not uniform for all members of the population. If no one was exposed, there would be no ''epidemic.'' If everyone were exposed, then either everyone would become victims (which is usually not the case), or an additional concept—differential susceptibility—would be needed to account for why some but not other persons contracted the illness.

Therefore, we can think of a two stage process. One is the *exposure* link representing contact between the source and the person. The other is the *susceptibility* link, representing the relation between the exposed person and the actual exhibiting of MPI symptoms. Each of these links depends on features of the situation and of the person. For any person in the situation, we can consider that he or she has two ''vulnerabilities.'' One is his or her vulnerability to being ''exposed'' to the ''contagious information.'' The other is his or her vulnerability to contracting the symptoms, given exposure. Each of these links merits some analysis.

Vulnerability to Exposure

There are a number of possible models of exposure. One is the idea that contact with a contaminant is random. Some people propose this view of who dies in combat. They say such things as ''if the bullet has your name on it . . . ,'' which presupposes that the probability of being hit with a bullet is random, at least with respect to the victim's actions. But a random model is probably not very viable for MPI. It certainly is not for combat.

If we posit a nonrandom model of who gets exposed we need to consider not only what substantive features predict differential exposure (sex, age, centrality in a social network, etc.). We also need to consider such things as *dosage level, frequency* of exposure and the like. Put another way: Does each contact have some probability of contaminating the person, with ''contamination'' being all or none (as in ''catching'' a germ)? Or, instead, does each contact represent an increment of contamination of the person (as in exposure to radiation)? If the latter, then we can pose two more questions: (1) Is there a ''threshold,'' above which the MPI symptoms occur and below which they don't? Or, instead, is the intensity of the symptoms a monotonic function of accumulated increments of exposure? (2) Does time mitigate effects of small amounts of exposure—either in a *time-decay* sense, where dosage rates need to be calculated ''per unit time,'' or in the innoculation sense, where small amounts provide a kind of immunization against later larger exposures?

Keep in mind that the "contaminating ingredient," in MPI, is *a complex set of information,* not a germ or a gamma particle. So when we talk about contact-contamination we are talking about: (1) Differential probabilities of different people in a situation being exposed to that critical set of information; (2) the probabilities that an exposure will be "received," psychologically; and (3) the effects of multiple exposures, including their incremental accumulation beyond some threshold and/or their potential innoculation effects.

When put in these terms, we are clearly talking about matters quite closely related to the whole communication, persuasion, and attitude change areas, for which there is a huge literature and a formidable body of theory. MPI research does not seem to have drawn upon that body of literature nearly as much as one would expect, given this analysis of the problem.

Vulnerability to Display of the Symptoms

Let us now consider what factors influence who exhibits the MPI symptoms, given exposure to the critical information at above-threshold levels. This is the question of differential susceptibility. Whereas exposure has to do with acquisition of information, susceptibility has to do with *acceptance* of the implications of that information and consequent behavior change.

The notion of susceptibility suggests its converse, resistance. We need to consider vulnerability as a continuum from highly susceptible to totally immune. A person who shows the symptoms is above some threshold on that continuum. What factors affect that person's vulnerability level? And what factors affect the threshold level, beyond which the MPI symptoms are displayed?

We have already considered some ways in which "dosage" levels and rates might serve either to increase cumulative dosages, thus adding to susceptibility, or to provide a kind of innoculation, thus reducing susceptibility. The communication-persuasion and attitude change research on amount of information, number, intensity and timing of inputs of information, and inocculation effects, should provide useful guides here. (See, for example, McGuire, 1969.)

There are other potential contingencies between exposure and susceptibility, too, and our MPI model should reflect them. For example, we could posit a model in which more exposure (i.e., more frequent, more intense, more sources of contamination, longer durations of contact) implies greater intensity of "dosage," hence greater susceptibility, therefore greater probability of showing the MPI symptoms. This would be a positive interaction between: (1) Amount of exposure, and (2) level of susceptibility of the individual.

We could, instead, posit a model that has a positive interaction between: (1) The number of others already affected (this could be considered intensity of exposure to the contaminating information) on the one hand; and (2) level of susceptibility, given exposure, on the other hand. In such a model, the more susceptible persons would become MPI victims first; then as more and more persons did so, individuals with more and more resistance would ''contract'' the symptoms. Such a model would have trouble accounting for why the MPI didn't eventually produce 100% victims. It would need either some basis for showing that as-yet-unaffected persons did not have sufficient exposure, or else some basis for showing that the unaffected had such high resistance that even the increased intensity, when all others had become affected, did not exceed their threshold.

Substantive Variables in Relation to Vulnerability

What substantive variables relate to exposure and susceptibility, and how do they tie into the network of variables involved in these potential models of vulnerability? In particular, how might the main consistent findings—about sex, stress, and somatization—fit into these models?

Sex Differences

The consistent finding that victims of MPI are mostly females could be viewed in several ways. It is *possible* that females have higher susceptibilities, either to all illnesses, or to ''hysteria'' type illnesses, or to ''imaginary'' illses. Such differences could arise from either biological/genetic bases or experiential/environmental bases. This is possible; but I don't think there is evidence to support it, nor do I think it is very likely.

Alternatively, females may be more vulnerable to *exposure*. They may gather and pay attention to rumors and gossip more than males; they may attend to broadcast media more than males. Again, I think this is unlikely on both theoretical and evidential grounds; but it is certainly a set of questions worthy of research.

Colligan and Murphy's (1978) attempts to account for sex differences on the basis of different sex role training implies that females are socialized more than males to express stress in somatic terms. So, given exposure (presumeably not sex dependent), and given high levels of stress (which may or may not vary by sex, as discussed in the following section), females are more *susceptible* to expressing their stress in MPI-like physical symptoms (and therefore are more likely to exhibit MPI symptoms) than males, who presumeably deal with their stress in other ways (including other physical symptoms, such as ulcers and coronary disease).

Colligan and Murphy's (1978) other explanation for the sex differences posits more sex-role conflict for the woman in the workplace. Here the hypothesis does not propose increased susceptibility of females to MPI symptoms given high stress and exposure. Rather, it hypothesizes that some females will have higher levels of stress because of role conflict. (See the following discussion of stress.)

None of the material I have covered has suggested that females may differ in how they acquire and process information. It would be well worth the MPI researcher's time to search for information on male-female differences both in the communication/persuasion and attitude-change literature, and in the literature of the currently booming areas of impression formation and information processing. As in so many other areas of social psychology, past research has tended to ignore sex differences. Sometimes there has even been a tendency to sweep them under the rug, on the grounds that they were not "theoretically interesting." So that search might be long and tedious; but it might pay off well.

Stress

There are several ways in which the stressful features of MPI situations might affect both exposure and susceptibility. For one thing, total intensity of stress on the person might act so as to sensitize the individual to possible "explanations." He or she might seek out information, and accept it more readily. On the other hand, stress may tend to narrow the focus of attention and reduce the individual's ability to take in and process information (See, for example: Appley & Trumbull, 1967; McGrath, 1970). The stressed person might become more sensitive to ways of coping with that stress. This could translate into higher susceptibility to MPI symptoms *if* that were a preferred coping style. Such a relationship would imply, however, that other people in the same high stress situations ought to show an increased use of *other* coping styles if they don't use the MPI symptoms. There does not seem to have been any attempts to test such a hypothesis. Such research surely seems worth doing.

Colligan and Murphy (1978) suggest that the stress stemming from poor worker-supervisor relations, and from poor conditions for worker-to-worker communication, deprive individuals of both normative and informational support that might counter the MPI-inducing information. Such poor communications do not reduce the person's exposure to the MPI-inducing batch of information (about the index case, later cases, and the symptoms) because these can flow through many channels including extra-setting mass media. But these poor communication conditions reduce the flow and the credibility of information that might offset the MPI. So the stress of poor communications may prevent reduction of vulnerability but not prevent exposure.

Somatization

One of the most interesting findings regarding MPI is that victims are persons who also show higher rates of absenteeism, sick leave days and medical consultations. Colligan and Murphy (1978) reviewed evidence indicating that this was probably *not* due to their being in poorer health in general. Their review leads them to suggest that somatization is a preferred coping strategy for MPI victims. Assuming this to be the case, then a preference for expressing stress in terms of physical symptoms can be considered an individual characteristic that increases susceptibility—given exposure and given high stress. Presumably, this characteristic does not affect exposure, except in the limited and indirect sense that frequent absences might reduce communication flow to that person within the setting.

An Attempt to Integrate

It seems appropriate to try to integrate all of this—the complexities, the cautions, the concepts—into a coherent formulation. It is probably appropriate to eskew the formal language of "real" science—assumptions, postulates, hypotheses, and all that—and offer a formulation stated as a series of questions and some plausible answers to them.

1. How Should We Regard an MPI? The most parsimonious formulation, it seems to me, is to regard an MPI as one of a number of possible alternative reactions to high levels of stress.

2. Where and When Do MPI Happen? MPI seem to occur *only* in work situations that have some or all of the following negative features: (1) Undesirable levels of conditions of the physical environment (e.g., noise, odors, temperature); (2) few positive-motivating factors in the job itself (e.g. monotonous, low pay, much production pressure); (3) unfavorable conditions of the social environment (e.g. poor communication, and friction) between workers and management; poor communication among workers, both because of physical impediments (e.g., noise) and because of other job features (e.g., assigned timing of work breaks). In other words, MPI occur in situations characterized by high levels of stress from the physical, task, and/or social environment of the work place.

3. Who Are Affected by MPI? Apparently, MPI occurs *only* for low status persons in the work situation. They *never* involve foreman or supervisors, much less high level managers. They always involve, exclusively, workers at the "production line" level (but, of course, not all of them). It should be noted that low status persons in a situation are probably under more stress than higher status persons in the same situations. They certainly have far fewer ways to cope with that stress.

4. When Is MPI Likely to be Used, Rather than Some Other Means for Coping with Stress? One of the ways a person can cope with high stress is by expressing bodily symptoms. This is probably a relatively preferred coping style for some individuals in any case. Those individuals, "somaticizers," are likely to be among the MPI victims if one occurs in a given situation.

But as stress increases, the array of possible coping options available to an individual tends to decrease. Similarly, as one's status in the situation is lower, one probably has fewer coping options. As the number of available coping options decreases, for whatever reasons, the probability of use of *any* remaining option increases, regardless of whether or not that option was originally a preferred coping style. Many coping options (quit the job; hit the boss; get drunk) tend to be eliminated as practical possibilities for individuals who are low in status and under high stress. But MPI—that is, the expression of stress in the form of somatic symptoms—remains always within the individual's response repertoire. Therefore, MPI will remain as a coping option as stress increases and other options slip away. It will *increasingly and inevitably become the coping method of choice,* even if it was not a preferred method at the start.

5. When Is a Particular Event Likely to Become a Triggering Event or Index Case for an MPI? A particular triggering event or index case is likely to serve as a "model" for an MPI if it is *noticeable, rational,* and *low-cost.* Triggering events are often noticeable because they represent changes in the work situation (e.g., a new work method, a new set of materials). Index cases are often dramatic because they involve dreaded diseases (e.g., polio). By "rational," I mean that the index case, the symptom set and the causal premise go together, are a "good fit," are a *plausible combination of cause and effect,* from the point of view of the victims (who believe that some toxic agent or pathogen is involved), regardless of whether they are considered plausible from the point of view of setting managers and medical practitioners (who believe that *no* toxic agent or pathogen is involved). The symptom set and alleged cause for individual MPI do, indeed, seem to fit this definition of rational, without exception. Indeed, it is precisely because the symptom-cause package of an MPI is highly plausible that it is so difficult to establish whether a given "outbreak" is an MPI rather than the result of chemical or biological agents, as discussed in the first section of this paper.

It also seems reasonable to suppose that a particular triggering event will be modeled as an MPI to the extent that it offers a (relatively) low cost means for coping with the stress. Consider what an MPI provides as a means for coping with high levels of stress in the work situation: (1) It can get the person out of the stressful situation, at least temporarily; (2) it can do so with no "blame" to the individual, and perhaps with little or no cost in lost wages, *provided* the "triggering event theory" of the cause is maintained; (3) moreover, it may lay some degree of blame on management, because most MPI in industrial settings involve alleged causal agents for which management bears at least some responsibility (e.g., a new solvent, a malfunctioning ventilating system); and, (4) expression of the MPI can represent evidence of one's empathy and solidarity with one's peers. Compared with hitting the boss (or the spouse), quitting the job, getting drunk, going over the boss's head, or getting an ulcer, expressing stress via bodily symptoms in relation to a dramatic and rational MPI triggering event certainly seems to offer a reward-cost net advantage! Perhaps we ought to ask, seriously, why MPI do not occur more often, and "infect" a far larger proportion of the work force when they do occur!

6. Why Do MPI Appear to be Almost Exclusively a "Womans Disease?" Probably for several reasons, but the most parsimonious formulation seems to me to be as follows: Women are low status in the work place. (So are the poor, the uneducated, the relatively young. So we would expect high MPI rates for them, too, and that seems to be the case.) Not only are women low status in any particular job setting, we would expect a very high proportion of women (and other low status categories, like poor and uneducated) to be in the "production line" jobs of the highly undesirable kinds of work settings in which MPI occur. So, we would expect a *very high proportion of women* (and other low status persons) *both as victims and as nonvictims in the settings where MPI occur*—and that indeed is what we find.

Since the higher the stress the fewer the coping options (and the more probable any remaining one, such as MPI); and since the lower the status the higher the stress and/or the fewer the coping options (hence the more probable any remaining one, such as MPI); it follows that persons who are low status in highly stressful situations—such as women in the highly stressful work settings characteristic of MPI—are *more likely to use* somatic symptoms (MPI) as a means for coping with that stress.

Note that this formulation "accounts for" *the high proportion of women who are not affected* in MPI settings, as well as *the high proportion of the affected who are women.* None of the other interpretations does so. Note, also, that the formulation presented here does not call upon "personality" constructs, nor differential somatization tendencies, nor differential sex role training. It offers a *sufficient* explanation in terms of stress, status and coping. Other explanations (e.g., differential role training; physiological or hormonal differences) may add to our understanding, but a very parsimonious explanation, in terms of stress, status and coping, is possible.

So, as an overall summary and concluding statement let me offer the following: MPI seem to be the response of some relatively low status persons in high stress work situations, subsequent to triggering events that are dramatic and that provide a rational and low-cost option for coping with that stress. Most MPI victims are women, but most *nonvictims* are also women, and the vast majority of women *do not* become MPI victims even in situations where MPI occur. Women are far more often to be found in low status jobs in high stress settings, where MPI occur. So, although MPI victims are invariably mostly women, MPI nonvictims are *also* invariably mostly women. What may be remarkable and worth study is that, in MPI type situations, *most* women cope with the high stress by methods other than MPI.

REFERENCE LIST

Appley, M. H., & Trumbull, R. *Psychological stress.* New York: Appleton-Century-Crofts, 1967.

Campbell, D. T., & Fiske, D. W. Convergent and discriminant validation by the multi-trait-multimethod matrix. *Psychological Bulletin,* 1959, *56,* 81–105.

Campbell, D. T., & Stanley, J. C. Experimental and quasi-experimental designs for research on teaching. In N. L. Gage (Ed.), *Handbook of research on teaching.* Chicago: Rand-McNally, 1963. (Also published as *Experimental and quasi-experimental designs for research.* Chicago: Rand-McNally, 1966.)

Colligan, M. J., & Murphy, L. R. Mass psychogenic illness in organizations: An overview. *Journal of Occupational Psychology,* 1978.

Cook, T. D., & Campbell, D. T. The design and conduct of quasi experiments and true experiments in field settings. In M. D. Dunnette (Ed.), *Handbook of Industrial and Organizational Psychology.* Chicago: Rand-McNally, 1976. (Also published as: Cook, T. D., & Campbell, D. T. *Quasi-experimentation: Designs and Analysis Issues for Field Settings.* Rand-McNally, 1979.

Kerckhoff, A. C., Back, K. W., & Miller, N. Sociometric patterns in hysterical contagion. *Sociometry,* 1965, *26,* 2–13.

Markush, R. E. Mental epidemics: A review of the old to prepare for the new. *Public Health Reviews,* 1973, *4*(2), 353–442.

McGrath, J. E. (Ed.), *Social and psychological factors in stress.* New York: Holt, Rinehart & Winston, 1970.

McGrath, J. E., & Altman, I. *Small Group Research.* New York: Holt, Rinehart & Winston, 1966.

McGuire, W. J. The nature of attitudes and attitude change. In G. Lindzey & E. Aronson (Eds.), *The handbook of social psychology* (Vol. 3). Reading, Mass.: Addison-Wesley, 1969.

Milgram, S., & Toch, H. Collective behavior: Crowds and social movements. In G. Lindzey & E. Aronson (Eds.), *The handbook of social psychology* (Vol. 4). Reading, Mass.: Addison-Wesley, 1969.

Runkel, P. J., & McGrath, J. E. *Research on human behavior.* N.Y.: Holt, Rinehart & Winston, 1972.

Sirois, F. Epidemic hysteria. *Acta Psychiatrica Scandinavica,* 1974, 252, 7–46.

Turner, R. H. Collective behavior. In R. E. L. Faris (Ed.), *Handbook of modern sociology.* Chicago: Rand-McNally, 1964.

Turner, R. H., & Killian, L. M. *Collective behavior.* Englewood Cliffs, N.J.: Prentice-Hall, 1957.

Webb, E. J., Campbell, D. T., Schwartz, R. D., & Sechrest, L. *Unobtrusive measures: Nonreactive research in the social sciences.* Skokie, Ill: Rand-McNally, 1966.

5 Mass Psychogenic Illness: Individual and Aggregate Data

Neal Schmitt
Michael Fitzgerald
Michigan State University

SUMMARY

At eight companies in which large numbers of employees became ill with no apparent physical cause, demographic, job satisfaction, and organizational climate data were collected. Both individual and aggregate company data on these variables were correlated with the number of physical symptoms reported by the employees. Results indicate a greater average number of reported symptoms in companies in which the total family income is low, there is dissatisfaction with the supervisor and personnel practices, and a perceived lack of staff support and clarity with respect to work assignments. It is argued that data concerning mass psychogenic illness are more appropriately analyzed at the company level. Problems associated with aggregation of data and guidelines as to how to proceed are discussed. Research involving multiple companies and assessing organization-level variables as well as individual variables is encouraged.

INTRODUCTION

The title of this chapter "Mass Psychogenic Illness: Individual and Aggregate Data" suggest that the phenomenon we are discussing is a group rather than an individual characteristic. In fact, in their review of published literature on this topic, Colligan & Murphy (1979) have defined contagious psychogenic illness as "the collective occurrence of a set of physical symptoms and related beliefs among two or more individuals in the absence of an identifiable pathogen." Whereas the *psychogenic* nature of the illness might be, and has been, debated, we shall concentrate on the group aspects of mass psychogenic illness and what this means in terms of research strategy and the interpretation of data collected to date.

Until recently, all research on mass psychogenic illness or industrial hysteria has been of a case study nature (see Colligan & Murphy, 1979, for a review). Whereas data were typically collected from many or all affected individuals, in all studies there was only one affected company or group of individuals. Case studies are certainly a rich and valuable method of inquiry especially in the early descriptive phases of studying a problem. However, in experimental or quasi-experimental work in which one desires to make generalizable statements, larger sample sizes are essential. Further, the point could be made that researchers engaged in studying mass psychogenic illness do not typically report their data in a way that indicates recognition of the essentially case study nature of their data. Treating instances of mass psychogenic illness as single events means that individual data from a company should be aggregated to allow computation of group scores. Hence, a series of methodological problems concerning aggregation and the appropriate unit of analysis arise. In this chapter, we will use data collected by NIOSH in eight different companies to illustrate some of these problems but also the reasonableness of treating companies (or work groups) as the unit of analysis in studying mass psychogenic illness. It should also be noted at the outset that the NIOSH data collection effort is the only one we have knowledge of which involved the systematic collection of similar psychological data in more than one affected unit.

RESULTS

The most commonly used research design has involved the comparison of affected and nonaffected individuals on a variety of demographic/psychological variables collected by interview or questionnaire in a company that has experienced a mass outbreak of illness for which there is no discernible physical cause. In this design, the sample size is the total number of respondents and the implication is that those variables that correlate with the affected status of the respondents are, at least associated with, if not determinants of the illness. Data of this type collected by NIOSH on selected variables thought to be associated with the symptoms of mass psychogenic illness are presented in the first nine columns of Table 5.1. These data are all self-report questionnaire data, the dependent variable being the number of symptoms out of a possible 25 reported by the subjects.

The first four variables are purely descriptive measures of employee tenure, hours at work, and income. The next set of fifteen items were single item scales measuring the degree of satisfaction the respondents reported concerning various aspects of their job situation. The last 10 variables were measured by the Work Environment Scale (Insel & Moos, 1974). Each of the 10 variables is measured by four dichotomously scored items. The names of these 10 scales are fairly descriptive of the items in the scales and are defined as follows:

> Involvement is the extent to which workers are concerned, committed to, and enthusiastic about their jobs.
>
> Peer Cohesion reflects the degree to which workers are friendly and supportive of each other.
>
> Staff Support reflects the extent to which management supports workers and encourages supportive relationships among workers.
>
> Autonomy is the degree to which workers are encouraged to be self-sufficient, make their own decisions, and develop their own potential.
>
> Task Orientation measures the degree to which the work climate emphasizes good planning and efficiency and encourages people to get work done.
>
> Work Pressure reflects the degree to which the pressures of deadlines and work dominates the climate.
>
> Clarity measures the extent to which workers know what to expect in their daily jobs and how explicitly rules and policies are communicated.
>
> Control is the degree to which management uses rules and pressure to keep workers under control.
>
> Innovation is the amount of variety and change which are emphasized in the work environment.
>
> Physical Comfort measures the extent to which the physical surroundings contribute to a pleasant work environment.

If one examines the correlations between symptoms and these 29 variables for the total sample (N = 867) there are a number of significant, but relatively small correlations with the number of reported physical symptoms. The highest of these correlations, that of Work Pressure and Symptoms, is only .182. Further, correlations across individual companies, are inconsistent; in many cases, the signs of the correlations are dissimilar. Some of this lack of consistency, or lack of significant correlations can be attributed to sampling error. Another reason lies in the possibility that there are really not any nonaffected people in these companies or that within a company everyone is similarly affected. If so, then there should be a restriction of range in the number of symptom variables and all correlations would be near zero. An examination of the average number of symptoms reported by the respondents in the eight companies indicated a range in the average number of symptoms from 1.49 to 5.44. A one-way analysis of variance indicated the companies were significantly different, $F(7,818) = 5.64$, $p < .0001$, in terms of the average number of symptoms reported by their employees. There are, then, significant company differences in the level of sickness reported by the respondents to the questionnaire.

Our next step in the data analysis was to take these symptoms averages and correlate them with averages on the hypothesized stressors. This, then, shifts the unit of analysis to the company level and leaves a sample size of 8. These correlations are presented in the last column of Table 5.1 and one can see at a

TABLE 5.1.
Individual and Aggregate Correlations of the Number of Symptoms with Various Stress Sources for Eight Instances of Mass Psychogenic Illness

Variable	*Columbus*[c] *N* = *91*	*Indiana* *N* = *94*	*Wisconsin* *N* = *65*	*Cleveland* *N* = *65*	*Aurora* *N* = *36*	*Tennessee* *N* = *229*	*Boston* *N* = *139*	*Huntington* *N* = *107*	*Total* *N* = *867*	*Aggregate r* *N* = *8*
Months of										
Employment	−.12	.06	.22*	−.06	−.29*	−.12	.00	−.22*	−.060	.03
Hours at work	−.06	−.04	−.15	.08	−.21	.01	−.05	−.10	−.004	.60
Personal Income	−.09	−.18*	.06	−.24	−.15	−.07	−.02	−.18*	.120*	−.55
Total Family Income	−.10	−.16	.31*	−.04	−.31*	.05	−.23*	−.17	−.12*	−.80*
Satisfaction Variables[a]										
Opportunity to use experience	.17	.13	−.13	.00	.15	.22*	.08	−.07	.068	−.46
Job Security	.16	.22*	−.12	−.07	.26*	.13*	.06	−.04	.109*	.44
Promotion System	.16	.17	−.16	−.03	.05	.19*	.08	−.07	.086*	.21
Overtime Pay	.19*	.10	.08	.15	.02	.05	−.13	−.13	.050	.65*
Excitement	.16	.06	.15	.02	.28	.09	.14	−.16	.034	−.53
Salary	.13	.14	−.14	.06	.08	.14	.01	−.03	.130*	.81*
Immediate Supervisor	.17*	.27*	−.18	.15	.03	.12*	.00	.01	.07*	−.24
Disciplinary System	.01	.23*	−.03	.07	−.04	.12*	−.01	−.11	.054	.72*
Frequency of rest breaks	.18	.04	−.16	.22*	.09	.04	.17*	.08	.058	.25
Work Scheduling	.11	.25*	.02	.11	−.01	.21*	−.03	.00	.120*	.57
Personal Appearance Code	.21*	−.11	−.13	−.09	−.19	.07	.19*	.02	.04	−.27
Determination of Days Off	.19*	.24*	−.12	.00	−.13	.18*	.06	.01	.13*	.77*
Performance Evaluation	.22*	.15	−.07	.11	.09	.25*	.02	−.04	.09*	.04
Freedom to make decisions	.05	.18	−.22*	.07	.19	.29*	.12	−.10	.11*	.02
Recognition from Supervisor	.24*	−.10	.08	.15	.00	.23*	.06	−.13	.06	−.27

Climate Variables[b]										
Involvement	−.19*	.00	−.07	−.06	.08	−.02	−.12	.01	−.045	.41
Peer Cohesion	−.16	−.21	−.11	−.20	−.03	−.01	−.05	.06	−.090*	−.58
Staff Support	−.26*	−.21*	−.14	−.29*	−.34*	−.14*	−.02	−.07	−.176*	−.65*
Autonomy	−.15	−.15	−.10	−.20*	−.22	−.10	−.17*	.00	−.102*	.28
Task Orientation	−.25*	.00	.06	−.14	.12	−.01	−.06	.23*	−.041	−.60
Work Pressure	.16	.36*	.04	.21*	.10	.23*	.01	.21*	.182*	.36
Clarity	−.31*	−.21*	.13	−.24*	−.31*	−.07	−.02	.10	−.110*	−.67*
Control	−.09	−.05	.07	−.11	.07	.03	−.13	−.13	−.14*	−.32
Innovation	−.21*	−.11	−.06	−.03	−.12	−.15	−.08	−.13	−.14*	−.52
Physical Comfort	−.24*	−.09	.00	−.18	−.35*	−.26*	−.18*	.10	−.172*	−.38
Climate Variables[b]										
Involvement	−.19*	.00	−.07	−.06	.08	−.02	−.12	.01	−.045	.41
Peer Cohesion	−.16	−.21	−.11	−.20	−.03	−.01	−.05	.06	−.090*	−.58
Staff Support	−.26*	−.21*	−.14	−.29*	−.34*	−.14*	−.02	−.07	−.176*	−.65*
Autonomy	−.15	−.15	−.10	−.20*	−.22	−.10	−.17*	.00	−.102*	.28
Task Orientation	−.25*	.00	06	−.14	.12	−.01	−.06	.23*	−.041	−.60
Work Pressure	.16	.36*	.04	.21*	.10	.23*	.01	.21*	.182*	.36
Clarity	−.31*	−.21*	.13	−.24*	−.31*	−.07	−.02	.10	−.110*	−.67*
Control	−.09	−.05	.07	−.11	.07	.03	−.13	−.13	−.14*	−.32
Innovation	−.21*	−.11	−.06	−.03	−.12	−.15	−.08	−.13	−.14*	−.52
Physical Comfort	−.24*	−.09	.00	−.18	−.35*	−.26*	−.18*	.10	−.172*	−.38

*$p < .05$

[a]These are all single item scales with a sixpoint response scale ranging from "Very Satisfied" to "Very Dissatisfied".

[b]These ten variables are 4 item scales developed by Insel and Moos.

[c]The N reported at the top of each column is the total number of respondents. Because of missing data, the sample size on which most of the correlations were computed was somewhat smaller.

glance that most are substantially higher than the nonaggregate data and 7 of the 29 correlations are statistically significant ($p < .05$). The direction and size of the significant correlations indicate that the companies whose employees reported higher personal incomes and particularly total family incomes had a lower number of average reported symptoms. Similarly, companies whose employees reported a greater number of symptoms had employees who expressed greater dissatisfaction with overtime pay, salary, the company disciplinary system and the determination of days off. There was a lower perceived degree of staff support and clarity of roles in these companies. Finally, there are a number of sizable, but nonsignificant correlations, all in the expected direction. This consistency is in marked contrast to the correlations based on single company data.

Our major thesis is that more studies of "mass" psychogenic illness or stress in general should treat the company or the work group as the unit of analysis. Before generating undue enthusiasm about the size of the aggregate correlations in Table 5.1, however, a number of cautionary statements are in order. First, it should be pointed out that all eight companies were affected and consequently, there should still be a restriction of range in the dependent variable and a resultant attenuation of the aggregate correlations reported in Table 5.1. This problem can be addressed adequately in future research and we will make some suggestions at the end of the chapter. The second problem involves analysis problems generated by the aggregation of data when indeed such aggregation is incorrect or, alternately, inappropriately drawing inferences about individual relationships based on aggregate data.

DISCUSSION

Aggregation Problems: Making Inferences about Individual Level Relationships from Aggregate Data

The comparison of data at different levels (company, work group and individual) produces aggregation problems. Robinson (1950) brought the "ecological fallacy" issue to the attention of social researchers forcefully when he concluded that one cannot use aggregated data to make inferences about individual correlations. He showed why the ecological correlation based on aggregated group data was likely to be higher than a similar correlation based on data collected from individuals except in the case in which individuals are randomly assigned to groups. That this is reasonable is clearly seen in Fig. 5.1 where the scatterplot for all individuals indicates a correlation of approximately .5. However, if one computes the correlation between x and y using group means, the correlation would be 1.00 in the case depicted, or at least higher than the correlation based on individuals. It should be noted that the correlation of x and y within any one

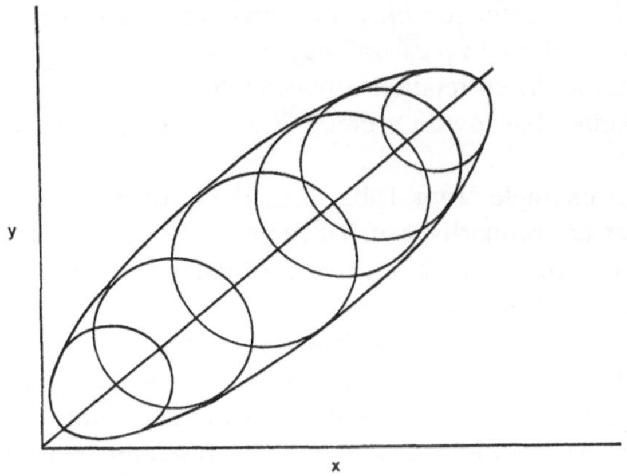

FIG. 5.1. Hypothetical situation depicting a moderate correlation between x and y on the individual level and a correlation of 1.00 between *x* and *y* for grouped data based on six subgroups.

company would be close to zero. This explains some of the difference between the total sample correlations and the aggregated correlations in Table 5.1. Hammond (1973) has identified this situation as one of ''homogeneous grouping'' and rightly indicated that it is not a serious problem because the regression coefficients for individual and grouped data are identical in this case. That this is true is also evident in Fig. 5.1 as the same straight line describes the relationship between *x* and *y* for individual and group data. Consequently, these regression coefficients are useful in drawing inferences about individuals from aggregate data. It should also be noted that Hammond, and sociologists in general, are usually concerned about inferences from the aggregate level to the individual when the individual level data are nonexistent or hard to retrieve, so the nature of these relationships is extremely important. This inability to get individual data may also occur in studies of mass psychogenic illness when, for example, investigators collect and correlate company turnover, absenteeism, or health statistics with personnel practices.

What Hammond (1973) labels aggregation bias, however, does present even more serious problems when one hopes to make inferences about individuals based on aggregated data. Aggregation bias occurs when the relationship between the variables of interest is systematically different in tbe different aggregate units. This is a specification error in that the variable defining the difference in aggregate units is not identified and included in the set of hypothesized relationships. These specification errors occur when deviations from equality of within group regressions are systematic. For example, if one is interested in the relationship between religion and voting patterns, and the density of people of one religion or another affects the proportion of that religion voting one way or the other, one has aggregation bias. In a precinct where 95% of the voters are Catholic, custom and group pressure may be more effective in producing a solid Democratic vote, so the correlation between religion and voting pattern would be high in that precinct but low in a precinct where the proportion of Catholics is smaller.

Or, using an example from Table 5.1, if the questionnaire respondents in some companies are primarily married females with large numbers of children there may be an artificially high positive correlation between total family income and the number of symptoms reported. This could happen because the phenomenon is more frequently associated with women and there is some indication that it is also associated with the strain arising from the financial and emotional responsibilities associated with a large family and the necessity to hold a job. This would mean that one should expect a positive correlation between the aggregate family income and the aggregate number of symptoms, when in fact this relationship has to do with response rates or the nature of the family status of the people in different companies.

Alternately, if there is a differing proportion of women respondents across companies and there is the typical sex difference in income, a spuriously high negative correlation between aggregate personal income and aggregate symptoms would be observed. This could and probably has been produced through differential pressure to respond to questionnaires. In one data collection effort by the senior author unrelated to mass psychogenic illness, the proportion of women respondents was twice as large as the male respondents. If nothing is done in some companies to encourage participation while participation in other companies is mandatory, the lower average income of females and their seemingly greater susceptibility to mass psychogenic illness would produce the observed high negative correlation between average personal income and average number of symptoms. The unspecified variable in this model would be the proportion of male versus female respondents.

A similar spurious elevation of the aggregate correlation between one of the job satisfaction variables and the number of symptoms reported in Table 5.1 is possible. If the determination of days off is more important to women employees whose family obligations are greater, then differences in proportion of women respondents across companies would serve to raise the aggregate correlation between dissatisfaction with determination of days off and symptoms. It should be noted that within company correlations for both the examples cited from Table 5.1 are quite different—with some negative and some positive. This, of course, is the difference among within group regressions which Hammond (1973) identified as a possible aggregation bias problem. Whether there is a problem or not depends on whether the company differences are systematic rather than random. Obviously, then one needs to know the phenomenon he/she is interested in before data collection in order to collect data on the appropriate variables. It is equally obvious that this is not always possible.

Statistical Evaluation of the Appropriateness of Aggregation

If one has data on the appropriate variables both at the individual level and the group level, there are a number of guides that indicate the likelihood of problems associated with cross-unit inferences and/or the appropriateness of aggregation. Recently, Burstein (1978) has evaluated the usefulness of three approaches to testing the appropriateness of grouping methods. Two of these methods—the structural equations approach (Hannan & Burstein, 1974) and the $\overline{X}$ - rule (Firebaugh, 1978)—proved adequate in estimating differences in individual and group coefficients when various types of aggregation bias were present.

The question of concern has to do with the degree to which the regression coefficient relating x and y at the individual level (β_{yx}) and the aggregate level ($\beta_{\overline{y}\overline{x}}$) diverge. Hannan and Burstein (1974) dealt with the problem by expanding the linear model to include the relation of the grouping variable to both X and Y. These latter relations are then used to forecast the consequences of grouping by any specific characteristic. Formally, the difference between the individual and aggregate coefficients is shown to be equal to:

$$D_{HB} = \beta_{yz.z}\beta_{xz}\sigma_z^2 \frac{\sigma_x - \sigma_{\bar{x}}^2}{\sigma_{\bar{x}}^2\, \sigma_{\bar{x}}^2}$$

where z represents an ordered grouping variable. This equation is instructive because it indicates the conditions under which there is no difference between grouped and ungrouped data namely,

(1) Z has no effect on Y when X is partialled out or $\beta_{yz.x} = 0$,
(2) Z has no effect on X or $\beta_{xz} = 0$, or
(3) $\sigma_{\bar{x}}^2 = \sigma_{\bar{x}}^2$.

When differences between grouped and ungrouped coefficients are observed, then some subset of the three conditions listed above does not hold.

The divergence of $\beta_{\overline{y}\overline{x}}$ and β_{yx} are estimated by Firebaugh (1978) in the following fashion:

$$D_F = \beta_{y\bar{x}.x}\,(1 - \eta_{\bar{x}}^2)$$

where $\eta_{\bar{x}}^2$ equals the correlation ratio of the between group to total sums of squares. This implies, as Firebaugh (1978) indicated, that downward cross-level inferences avoid bias when the group mean is unrelated to Y and X is partialled from Y.

Both of these approaches are useful in the single predictor case, but they become cumbersome and impossible as the number of regressors increase and/or the grouping variable is a nominal one. Most importantly, these techniques are useful only when one has both grouped and ungrouped data; when this is true, however, one can compare the coefficients directly. Such procedures might be useful when it is too expensive to analyze all the data, though a random sample of the cases would also be appropriate. In some instances, disclosure of individual cases is unethical. When this is true, one could use the procedures above to assess the seriousness of the distortion resulting from aggregation; if there are no differences one could report aggregate data and draw inferences at the individual level.

One can also test for the significance of the difference between grouped and individual coefficients using Burstein's modification (1978) of the Fiege-Watts statistic (1972):

$$F = \frac{(\beta_{\bar{y}\bar{x}} - \beta_{yx})^2 \left[\frac{1}{SS(\bar{x})} - \frac{1}{SS(x)} \right]^{-1}}{SS(res)\ /N-1}$$

SS($\bar{x}$) and SS(x) are the sums of squares for X for aggregate and individual data respectively;
SS(res) equals the sum of squared residuals from the individual-level regression model; and
N is the number of individuals.

Like other significance tests, this test is dependent on the sample size and the investigator is still forced to make decisions about what difference between coefficients is practically important.

Where does this discussion lead in terms of practical guidelines when aggregation bias is possible? First, one should heed Burstein's guides for collecting and maintaining data; namely, (1) measure all variables at their lowest possible level; (2) data from individuals should be matched with data from the groups to which they belong; (3) all information should be identifiable at its lowest possible level with retrieval capabilities at multiple levels; (4) use whenever possible, measures with established reliability and validity. Further, any decision to aggregate individual data should be based on sound conceptual grounds. This means that data on grouped variables which might affect within group relationships must be collected. Fortunately or unfortunately, we have no better guide concerning what data to collect than that suggested by our theory(ies) and understanding of the phenomena of interest.

Aggregation Problems: Deciding When to Aggregate Individual-Level Data

Thus far the discussion of aggregation bias and/or the ecological fallacy has focused on the distortion which may occur when one attempts to infer something about the relationships among variables for individuals when one has only group data. Another question involves appropriateness of the aggregation of individual level data (for example, the attitude and demographic variables in Table 5.1) to get group scores which are then related to inherently group data (for example, company size or technology). Again, it must make theoretical or conceptual sense to say that these attitudes/participation variables are characteristics of the company. For example, differences in the clarity of articulated personnel policies would result in company differences in the job satisfaction and climate variables, hence they might justifiably be aggregated. Beyer (1978) has suggested that a minimal test of whether it makes sense to treat individual data as a group property is whether the ratio of the variance of responses between groups is significantly larger than the variance of responses within organizations. If this F-ratio is significant, we have support for the existence of a particular construct as a property of the groups in question. For all the variables reported in Table 5.1, this F-ratio was statistically significant ($p < .05$).

If the F-ratio is not significant, the construct may apply only to the individual level of analysis or the investigator may have erred in its operationalization. A nonsignificant F-ratio indicates lack of within group consensus concerning the variable in question, or in a psychometric sense, a lack of respondent consistency. When aggregation seems reasonable conceptually but there is no unit consensus, a careful rewording of items may indicate to the respondent what level of analysis one is interested in. In this context, it may be better to ask questions that one is interested in aggregating in a descriptive sense rather than an evaluative one. For example, it would seem better to ask people how often they offer suggestions to management about work procedures than how they feel about participating in their company's activities.

In summary, two variables that are to be correlated must be at *conceptually* similar levels of analysis. The decision to deal with aggregate or individual data makes sense only when it is relevant to the question of interest. A theoretical base for one's investigation or at least an educated guess as to the relevant variables affecting the phenomenon of interest will dictate the level of abstractness. In the case of psychogenic illness, studies of organizational-level stressors using organizations as subjects would seem particularly appropriate.

Future Research

In first suggesting the study of psychogenic illness at the organization level, we noted that the data presented in Table 5.1 are all data from "sick" companies; consequently, company scores on the symptoms variables as well as its correlates will be elevated and there is likely some restriction of range. However, there were still significant company differences. How does one investigate the correlates or determinants of mass psychogenic illness as occurrences are quite rare and we certainly don't know when to predict an instance of the phenomenon? Further, illness affects the variables measured and one must collect data in an emotionally charged atmosphere.

If mass psychogenic illness is viewed as the final stage in an organization's reaction to stress, then we should be able to study random samples of nonaffected organizations to achieve insight into organizational stress. Companies that experience outbreaks were likely under stress for a considerable period of time and are probably not much different from many other companies in which stress and the reaction to stress is present but less dramatic. Whereas generalizations to the organizational phenomenon are obviously dangerous, those familiar with Selye's work (1974) will note that this idea is similar to his description of the general adaptation syndrome in various species. That is, individuals can carry on relatively normal lives under extreme stress for long periods of time before the final exhaustion stage occurs. We are proposing that an outbreak of mass psychogenic illness is similar to an individual's exhaustion stage. This implies that one does not need to study only affected companies, but that one could study a random sample of organizations to gain an understanding of the determinants of mass psychogenic illness.

What variables should one study as possible indicants of a stress-prone organization? Two recent conceptual reviews (Beehr & Newman, 1978; McGrath, 1976) of the stress literature as well as the data in Table 5.1 suggest some possible correlates of mass psychogenic illness. McGrath suggests three major sources of stress: the physical technical environment, the social interpersonal environment and the person. Each of these components also interacts to produce additional sets of variables. For example, a person entering a given social context gradually develops a role. The inability to understand how one fits in, or role ambiguity, has been identified as a significant source of stress in several previous studies (for example, see Rizzo, House, & Lirtzman, 1970, and Table 5.1.). Similarly, Beehr and Newman (1978) indicate a set of possible environmental stressors and a number of person characteristics thought to be associated with stress and suggest that there are also possible person-environment interactions.

Specifically, Beehr and Newman list such organizational characteristics and conditions as company size, job security, union constraints, required relocation, organizational structure, work hours, staffing policies, management style, evaluation, control and reward systems, sociotechnical changes, training programs, advancement opportunities, and subsystem relations. Of this list, Beehr and Newman (1978) indicate that only company size has been related to stress. The data in Table 5.1 should encourage further study of the role of management and staff in creating stress. Role conflict and role ambiguity as determinants of stress have been frequently studied, but to our knowledge not as organization-level variables. We have no data on the degree to which the organization's external demands or conditions produce stress. We have no data relating stress to location of the work place and the availability of decent transportation, number and nature of clients and customers, union strength, government regulation, technological and scientific developments, suppliers of materials and services, and consumer movements. There is some information on the effect of job demands and task characteristics; for example, the effects of shift work have received some attention (Rentos & Shepard, 1976; Tasto, Colligan, Skjei, & Polly, 1978). Beehr and Newman indicate there is no available information concerning responsibility levels or the job characteristics hypothesized to increase the intrinsic interest level of the work. Further, the economic problems or instability of the company or industry may serve to produce stress on company executives which in turn produces stress for others in the organization. Finally, there is no reason why a study indexing these psychological/organizational variables should not also include systematic collection of data on the use of chemicals or possible toxins by the organizations.

All of these variables could be conceptualized as organization-level constructs and all could reasonably be expected to contribute to the stress experienced by a group of people. Relative to the number of potentially important elements listed by both McGrath (1976) and Beehr and Newman (1978), the number of elements that have been studied is few.

We think that a study or series of studies comparing stress levels in various organizations to the organizational variables suggested above would be extremely valuable in increasing our understanding of stress in general and mass psychogenic illness in particular. Such studies should not rely solely on questionnaire measures; in fact, many of the variables suggested above could not be measured in this way. Finally, these studies should include the collection of longitudinal data and the use of "natural changes" to allow for stronger causal statements about the factors involved in stress.

REFERENCES

Beehr, T. A., & Newman, J. E. Job stress, employee health, and organizational effectiveness: A facet analysis, model, and literature review. *Personnel Psychology,* 1978, *31,* 665–700.

Beyer, J. M. *The use of aggregation in measuring structural properties of organizations: A sociological point of view.* Paper presented at the Annual Convention of the American Psychological Association, Toronto, September 1978.

Burstein, L. Assessing differences between grouped and individual-level regression coefficients. *Sociological Methods and Research,* 1978, *7,* 5–28.

Colligan, M. J., & Murphy, L. R. Mass psychogenic illness in industry: An overview. *Journal of Occupational Psychology,* 1979, *52,* 77–90.

Fiege, E., & Watts, H. W. An investigation of the consequences of partial aggregation of microeconomic data. *Econometrica,* 1972, *40,* 343–360.

Firebaugh, G. A rule for inferring individual-level relationships from aggregated data. *American Sociological Review,* 1978, *43,* 557–572.

Hammond, J. L. Two sources of error in ecological correlations. *American Sociological Review,* 1973, *38,* 764–777.

Hannan, T., & Burstein, L. Estimation from grouped observations. *American Journal of Sociology,* 1974, *39,* 374–392.

Insel, P. M., & Moos, R. H. Psychological environments: Expanding the scope of human ecology. *American Psychologist,* 1974, *29* 179–188.

McGrath, J. E. Stress and behavior in organizations. In M. D. Dunnette (Ed.) *Handbook of industrial and organizational psychology*. Chicago: Rand-McNally, 1976.

Rentos, P. G., & Shepard, R. D. (Eds.) *Shiftwork and health: A symposium*. U.S. Dept. of Health, Education, and Welfare, Washington, D.C., 1976.

Rizzo, J. R., House, R. J., & Lirtzman, S. I. Role conflict and ambiguity in complex organizations. *Administrative Science Quarterly,* 1970, *15,* 150–163.

Robinson, W. S. Ecological correlations and the behavior of individuals. *American Sociological Review,* 1950, *15,* 351–357.

Selye, H. *Stress without distress*. Lippincott: New York, 1974.

Tasto, D. L., Colligan, M. J., Skjei, E. W., & Polly, S. J. *Health consequences of shift work*. Cincinnati, National Institute of Occupational Safety and Health, 1978.

6 Evidence and Inference in Research on Mass Psychogenic Illness

Robert F. Boruch
Northwestern University

INTRODUCTION

At one point in his illustrious career, Niels Bohr announced that "Every sentence I utter must be understood not as an affirmation but as a question." It is tempting to espouse this theme in a paper on mass psychogenic illness. For what is known about the phenomena is rudimentary and often ambiguous. Moreover, I am no expert on the topic. One ought to be a little suspicious that such denial attempts to deter criticism. So I do claim expertise in examining evidence on social problems and on purported solutions.

This chapter's main purpose is to discuss the quality of evidence and inference in research on mass psychogenic illness. The resources exploited here are published case studies of the phenomena, and the excellent reviews by Sirois and Markush. These include cases of illness in schools, factories, and hospitals, in the U.S. and abroad. Though the emphasis here is on criticism, I do not mean this to be adverse commentary.

Rather, my intention is to identify what is admirable practice and what is not, to clarify evidential problems that have not been resolved, and to offer suggestions for their resolution. The scientific muses invoked generally concern access to information, especially for the sake of competing analysis, quality of evidence, and reasonableness of inference.

Colligan and Murphy's (1979) definition of mass psychogenic illness is adopted here: the collective occurrence of physical symptoms and related beliefs among two or more persons in the absence of an identifiable pathogen. More formal definition, considered later, would attend to the ideas that onset is rapid, affecting several people in a short time period, that effects are temporary, and that symptoms of affected people are similar. Laughlin's theory of conversion is also adopted as a vehicle for understanding the phenomena at the individual level (Kutash, 1965). His conception is that conversion is an unconscious process by which intrapsychic conflicts that might give rise to anxiety instead secure symbolic external expression. The adoption of definitions is tentative. Most readers will recognize better than I that they are obtuse, and one of the purposes of this examination is to better understand how they might be improved.

SYMPTOMOLOGY AND MEASUREMENT

Several features of reporting on symptom patterns make it unnecessarily difficult for the conscientious reader to understand the phenomena. These include published descriptions of the uniformity, verifiability, and recurrence of symptoms, and the influence exercised by observation itself on reported symptoms.

Uniformity

Ebrahim (1968), for instance, suggests, that each of the 300 East Africans discussed in his case study were afflicted with running mania. But it is not clear whether each exhibited symptoms to the same extent, whether each also exhibited other symptoms typically, or whether there were remarkable or atypical cases. Whether all cases were afflicted uniformly by headaches, burning eyes, nausea, weakness, etc., in Colligan and Stockton's (1978) study of a garment manufacturing plant, Stahl and Lebedun's (1974) work in the data computation center, and Chew, Phoon, and Mae-Lim's (1976) investigation of the Singapore factory, is similarly ambiguous. Knight, Friedman, and Suliante (1976), on the other hand, provide evidence that uniformity in some respects and heterogeneity in others ought to be expected. Hyperventilation preceded the salient symptom, fainting, among some girls in the Louisiana school episode. Headache and dizziness preceded. Behavior akin to petit mal or catatonic posturing preceded still others. Colligan and Murphy (1979) manage to provide the most information by condensing frequency of symptoms encountered in an electronics plant episode, but did not take the next natural step to identify clusters of symptoms.

The implication then is that uniformity ought to be regarded as a legitimate topic of investigation during an episode and the regularity or lack of it ought to be reported. Neither action appears to be common now, judging from published reports. How useful such information can be depends on environmental insult. I understand from Sax (1979) that some toxic substances, such as lead, produce a variety of symptoms depending on the individual who ingests it, but other substances do generate uniform symptoms. Consequently, a specific number of symptoms and regularity in their appearance is unlikely to be an adequate indicator of mass psychogenic illness in all cases. Examining cases on the basis of uniformity is still warranted if we want to establish utility of the information.

Relapse

Whether symptoms reappear spontaneously still appears to be debatable, perhaps because so few follow-up studies have been undertaken. Sirois (1974), for instance, thoughtfully examines 78 cases of mass psychogenic illness and observes that "... relapses in the same person are not the rule [p. 15]." He is a bit inconsistent in that his tables show that 19 out of 51 accessible cases involve relapses. That relapse does occur, that there may be multiple relapses, and that it is possible to document them reasonably well is illustrated by the Chew et al. (1976) work in Singapore. Phoon's remarks in this volume suggest that reoccurrence of episodes in certain Singapore factories is not uncommon.

A more complicated problem concerns an individual's relapse following attempts to investigate or treat the problem. The return visit of Knight, Friedman, and Sulianti (1965) to the Louisiana school in which girls blacked out in a mass episode was followed quickly by other blackouts. It is as reasonable to regard some of these as relapses as to regard them as a product of the intervention, a problem considered in a bit more detail later.

The obvious implication is that one ought to monitor systematically for relapse among individuals in a group previously affected. The second, less obvious, implication concerns relapses which are plausibly caused by visits of the investigator, as in the Knight et al. case, or by attempts at treatment, as in the boss's visit to hospitalized women in the Colligan-Stockton (1978) work on a garment manufacturing plant. In particular, if such visits do indeed provoke relapse, and pathogens, toxic chemicals, allergies and the like are not present during visits, then it becomes more reasonable to diagnose at least the relapse as psychogenic and less reasonable to attribute the problem to environment. Indeed, the diagnostician may wish to introduce such visits deliberately in the interest of crude and cheap test of the hypothesis that the illness is psychogenic. A third implication is that relapse rates deserve attention because they are part of symptomology. Because they help to identify the disease entity, and to build theory on immunity and sensitivity, and to identify the effects of ameliorative action, documenting rates is essential.

Irrelevant Variation in Observed Symptoms

The quality of information about symptoms is influenced by the way one makes observations and by the setting itself. Factors that can plausibly degrade quality are not often recognized explicitly in published reports. Their appearance in other kinds of action research suggests that they ought to be at least considered for investigating mass psychogenic illness.

It is possible, for example, that the mere observation of an individual's behavior will produce real changes in that behavior. So, for example, Knight, Friedman, and Sulianti's (1965) report on blackouts among Louisiana school girls contains a nice chart to illustrate the incidence of attacks following a quiet period and the reentry of investigators to the school. The notion becomes more plausible when we recognize arguments that similar inflation of symptoms occur with other types of medical examination. So-called labile hypertension, for example, registered by temporarily elevated blood pressure, may be partly a function of nervousness produced by a visit to the physician's office (see Kolata, 1979).

The Knight et al. (1965) case is the only published illustration of the sort I've been able to find. The problem *may* have occurred in other cases of mass psychogenic illness, e.g., the Singapore factory episodes described by Phoon in this volume. But pertinent data do not appear to have been published. What makes the matter more intriguing is that related problems are *alleged* to appear in other action research. For example, Kerckhoff, Back, and Miller's (1965) June Bug investigation was said to have been delayed two months because plant managers were apprehensive about the possible effect of interviews. School districts have refused to permit sociologists to ask students questions about drug abuse on account of fear that such questions would generate an interest in using drugs. Desegregation researchers have been asked to terminate studies on grounds that squabbles would be produced by their merely asking questions. Such claims are occasionally self-serving; consequently they may be entirely misleading. But we know so little about the way investigation affects behavior in disruptive settings that we cannot afford to ignore the matter.

It is also true that, irrespective of behavior, the information recorded during an investigation can be influenced by the views of the investigator. Knight, Friedman, and Sulianti (1965), for example, report that psychiatric interviews were informative. But from their text, it appears that interviews were not blind. The lack of blindness and psychiatrists' preconceptions could have led to the judgments that victims of attack were more "sexual," more willing to talk about illness, and so on, than were unaffected students. The problem, of course, is not confined to investigation of mass psychogenic illness. Blind interviews and ratings are a natural device for avoiding biases attributable to prior views of interviewer and rater. The problems of assuring blindness are mechanical, but I should think that it is feasible to obtain blind observations in interviews if topics are prescribed and some topics are temporarily proscribed.

The Effect of the Setting and the Episode on Test Results

The unstable settings in which interviews take place can affect the quality of observations. Knight et al. find no interesting results from their Rorschach testing, for example. Assuming that such tests are informative in ideal settings, an assumption I find unattractive, the failure to find anything interesting may be attributable to low reliability of such tests in disruptive settings. Similarly, the extraneous influences on the Minnesota Multiphasic Inventory, perhaps electroencephalograms, and other measures reduce the likelihood of interesting differences between an affected group and an unaffected one. A failure to find such differences doesn't mean there are none. It may mean that test results are insufficiently reliable.

The possibility that test results will be degraded by random variation implies that conditions ought to be controlled so as to minimize disturbance, e.g., testing an individual in isolation. Regardless of whether testing conditions can be controlled, the conditions of testing ought to be reported: They are a real part of the test itself. Further, estimates of internal consistency and related forms of reliability for psychological inventories, rating forms, and physical measures, and estimates of test-retest stability in the interest of understanding temporal reliability of measures are warranted. Both types of statistic, internal consistency and stability, can help to determine whether further analyses are justified. There is not much point to even rudimentary comparisons of, say, affected and unaffected groups if the probability of finding differences is degraded drastically by random noise in the observations.

In this volume, Dr. Markush points out that it is also possible that an episode of mass psychogenic illness itself may affect test results systematically. For example, an individual's score on the hysteria scale may be inflated following an episode. Inferences about the personality of individuals who are susceptible to attack would then be misleading: The score is high because of attack, not necessarily because the trait was present in greater degree prior to the event. If indeed, measured personality dimensions are not robust against situational disturbance, if the trait is not as durable as IQ say, then this problem is indeed severe. Establishing susceptibility would be virtually impossible. On the other hand, knowing that disruption influences scores on such inventories, if that is in fact what occurs, is itself scientifically interesting.

It is not unreasonable to entertain a similar theory for social measures, especially if observations are made well after an actual episode of mass psychogenic illness. For instance, the Kerckhoff et al. (1965) data on sociometric patterns appears to have been collected some two months after the June Bug episode. It is possible that friendship networks found at that point were influenced by the attack. That is, individuals who had been affected might feel some additional kinship as a result, and some friendships may evolve or be strengthened as a consequence of common experience. If so, inferences about the role of social networks in mass psychogenic illness will be misleading.

In the absence of any other information, several options are worth considering. It may, for example, be possible to select inventories on the basis of their robustness against extraneous influence such as an episode of psychogenic illness. The criteria for selection should be empirical, and this suggests a program of formal experimentation on the influence of high and low anxiety situations on scores. Some work is possible in the laboratory, and given the advances in the state-of-the-art in social experimentation, some research in the field may also be possible (Riecken & Boruch, 1974). The generalizability theory available for design of such experiments is elaborate (Cronbach, Gleser, Nanda, & Rajaratham, 1972).

A weak option is to develop new tests which in theory provide accurate estimates of the level of traits despite environmental variations. For instance, so-called Rasch models might be pertinent in that tests constructed in accord with the model may yield better estimates of trait level independent of the systematic pressures engendered by a disruptive setting. For applications of the model in achievement testing, where the objective is to estimate true ability independent of item difficulty, see Lord and Novick (1968). The method requires large samples.

A third option hinges on the idea that measuring the level of a trait *after* an event is useless at best and misleading at worst. Comparing level of hysteria to a norm is unlikely to be helpful, for example, because there are so many unknown and unknowable differences between any norm sample and the group actually involved in an episode of mass psychogenic illness. Measurements taken before an episode, in the interest of establishing individual susceptibility, are more desirable, but of course no such measures are typically available. The implication is that routine monitoring systems be set up to obtain such information regularly. Implementing such a system would be difficult. But if the objective is to predict outbreaks, to understand the effect of outbreaks, and to assay susceptibility without formal side experiments of the sort described earlier, I see no other way to meet the objectives.

A fourth option is applicable at least to social network inventories and is deceptively simple. One may elicit retrospective reports about friendship patterns before the episode, for instance, as well as reports about contemporary friendships when delay in eliciting information is unavoidable. The problem here, of course, is that the quality of retrospective reports may not be sufficient to justify analysis. It is reasonable to develop a rough idea about the validity of such data from other investigations in which it is possible to obtain friendship information during *and* well after the episode. A discrepancy would be presumptive evidence on the degree to which memory failure, the attack itself, and other factors influence the accuracy of retrospective reports. And such evidence is invaluable to the investigator who must rely solely on the latter.

It would be presumptuous to declare that problems of the sort described here are important in every case but not presumptuous to regard the problems as plausible sources of irrelevant variation in the data. It is reasonable to ask that the problems be considered in published reports, even if they've not been resolved.

Moreover, identification or development of inventories and interview methods which are robust against environmental disruption require a program of research. There is fine precedent for methodological investigations of field surveys at the U.S. Census Bureau, the National Center for Health Statistics, and elsewhere.

Cooperative university-based laboratory research in the same spirit is worth considering to understand how current observation schemes can be improved.

Methods for Research in Unsettled Situations

Several writers express concern about the difficulty of doing high quality research amid the chaos characteristic of mass psychogenic illness. There are difficulties in gaining admission to the organization, in interviewing individuals upset by attack, and in interviewing nonvictims who may view victimization research with some hostility. Time for the research is usually scarce. There may also be resistance to research in general and psychology or psychiatry in particular.

It is puzzling that such difficulties and their consequences are not better documented because they affect the quality of information collected about the outbreak. The Colligan, Urtes, Wisseman, Rosensteel, and Anania paper (in press), an exception in this respect, is discussed later. There is small comfort in knowing that similar problems are poorly documented in social research on prison riots, campus protests, picket lines and unions, street gang violence, desegregation riots, natural disasters, and other disruptions. The difficulty may be dramatic in these instances, and it is clearly chronic (Beals, 1969; Boruch & Cecil, 1979).

Consequently, we are justified in regarding such difficulties as a legitimate arena for methodological research and development. For mass psychogenic illness, it is not unreasonable to consolidate the experience of gaining access through personal contact with managers, developing rapport through conventional devices, minimizing the time demanded of respondents, minimizing the provocativeness of inquiry, accommodating typical criticism, and so on. Developing a reward system, including paying victims and especially nonvictims, to compensate for their participation in their research is likely to help in this arena of research as it does in others. It is doubtful that science or methodology will be served well by ignoring the opportunity to advance the state of the art in this arena. But I know of no concerted effort to do so, or to link understanding here to what we know from research efforts in other unsettled situations.

MEDICAL AND ENVIRONMENTAL TESTING

Reporting Results of Medical Testing

Early investigators report that medical tests were undertaken on at least some victims of psychogenic illness. But they rarely report the specific tests nor do they cite reports containing test results. So, for example, Knight et al. (1965) report that some students "were subjected to thorough physical examination, including lumbar puncture [p. 858].'' But the authors of this fine report still fail to specify the tests and the number of students involved in each. Similarly, Stahl and Lebedun (1974) refer to ''excessive blood and urine tests on women'' in the computation center case, but provide no detail.

Reporting such tests in journal articles or research reports is important for several reasons. First, if the tests were pertinent enough, in some physician's judgment, to have been done, they are pertinent enough to be reported. We know what causes of symptoms can be ruled out of speculation, and we narrow the territory of the unknown. The second reason for reporting lies in the idea that we should have a better understanding of why we follow false leads, why our initial investigations or tests may be misleading, unnecessary, or silly. Understanding the traps in simple testing for gonorrhea, for example, led Mausner and Gezon (1967) to avoid relying on vaginal smears alone and ultimately to their development of a remarkable case study. The third reason for more adequate reporting of medical results is that such data can be important in secondary analysis by independent researchers. If at Time 1, malady A is ruled out by such tests in early investigations, one might rule out malady B based on the same test or regard B as a possibility based on the test and new information in later investigations.

Reporting Environmental Tests

Though many analysts report that environmental tests were done as a consequence of an outbreak, few report the specific nature of tests or cite original reports on the topic if they exist. For example, despite the attention paid by Stahl and Lebedun (1974) to noise of a diesel engine in their case study of the data computation center, no detail on the nature of noise or tests for gas is provided. In other cases, claims that ''samples of ambient air in different quarters, of solvents, chemicals, and glues, even of drinking water and food in vending machines'' are not supported by reference to specific tests. The reasons these tests ought to be specified, despite their negative results, are identical to those cataloged earlier for medical reporting. They include: ruling out competing explanations and limiting the unknown in the case at hand; identifying plausible explanations for an epidemic which are incorrect; and permitting later secondary analysis of the case.

Herrick (1979) reports that the reliability of air screening is such that readings are within ±25% accuracy. This suggests that reports of environmental tests should routinely provide information about their reliability, just as one ought to provide estimates of reliability of personality inventories, questionnaires, and the like. The reliability of environmental tests no doubt also depends on circumstances, so an experimental testing program for such tests, analogous to one outlined earlier for softer measures, is warranted, especially if estimates cannot be obtained during an outbreak.

Environmental Sampling

Very few reports in the social science literature provide any real information on the environmental *sample designs* employed during investigations. By "sample design" here I mean the plan that specifies when, where, and in what volumes the units will be sampled. I do not mean what appears to be a vernacular definition of sampling technique used by Herrick (1979), i.e., the physical device used to capture the unit of gas, liquid, or solid. I expect that most samples are actually opportunistic, e.g., someone points out an area in which a peculiar odor was detected and the area is examined once or twice. Apparently, obtaining a sample likely to exhibit extreme properties, in the interest of upper bound estimates, is not uncommon, e.g., a rafter sample for certain precipitants. Perhaps this is the only way one can sample in such cases. But to label an event as "taking a sample" without further specification is, at best, frustrating to scientific readers. The word sample is used promiscuously and we become suspicious in the absence of additional information.

More important perhaps is the absence of interesting work on statistical sampling plans for such cases. I can find no textbooks or guidelines that help answer questions such as: How often should one take a sample? What sample design ought to be exploited for assessing physical areas, or volumes of moving air? What kinds of systematic biases in sampling or in testing a sample at any given time can be expected? Answering such questions is reasonable given that odors emerge and dissipate, noise varies in intensity, and so on, and static tests are unlikely to be very useful.

DIAGNOSIS AND MISDIAGNOSIS

Simple Theory of Diagnosis

Implicit in Sirois' (1974) remarkable review of epidemic hysteria is a theory of diagnosis and misdiagnosis. In the abstract, a theory of this sort is important for any social or medical research. In practice, no explicit theory is treated or exploited in reports on mass psychogenic illness despite the available technical resources (Cochran 1968; Murray, 1971).

Sirois' simple theory implies that establishing the nature of the phenomena entails recognizing that similar symptoms may generate two different forms of explanation. That is, the diagnosis may be that the problem is largely medical in origin, i.e., there are toxic influences on physical well-being. Or, the diagnosis may be that the illness is, in this case, contagious hysteria. Either declaration might be wrong in interesting ways.

For instance, Sirois cites the McEvedy and Beard (1973) research to support the notion that some disease is based more on physicians' expectations than on biomedical evidence. Recall that in their study of neuromyasthenia, symptoms similar to those exhibited by victims of poliomyelitis were also exhibited by individuals without the latter disease, and were regarded as a clinical syndrome when indeed the problem appears to have been psychogenic. Barnes' (1977) review of worthless surgery is as pertinent. He reminds us that ptosis was characterized early in this century as a condition in which the position of internal organs was "abnormal." The abnormality thought by surgeons to cause a wide variety of symptoms, e.g., kidney displacement, was alleged to produce neuroticism, back pain, and vomiting. We know now that ptosis is not an organic problem, that surgery was unwarranted, and that diagnosis and etiology were nonsense.

This class of misdiagnosis implies that we ought to catalog stereotypical errors in medical diagnosis. The complement of the rate of accurate medical diagnosis includes a rate of psychogenic illness and the probably larger rate of misdiagnosed medical problems. Crude estimates might be based on contemporary research in physicians' second opinions by the National Center for Health Services Research, and perhaps on contemporary archives, e.g., insurance company records.

If the declared diagnosis in a particular case is contagious hysteria, this too may be wrong in interesting ways. Young women, for instance, occasionally complain that their physicians regard some of their symptoms as hysterical when, the women maintain, the problem is biomedical. Sirois (1974) relies on more reliable data to argue that epilepsy and rheumatic fever are examples of disease which are triggered by or mimic collective symptoms. Recall also the reminders in this volume that diagnoses of acute schizophrenic reactions among textile workers have been made when in fact the origin of the symptoms is psychogenic but not tied directly to hysterical conversion.

This implies that one ought to catalog chemicals, gases, and so forth partly on the basis of whether they produce symptoms similar or identical to those encountered in mass psychogenic illness. Sirois observes that the latter is benign in having no permanent effects. It should then be possible to classify toxic agents partly on the basis of temporariness of their effects, diffuseness of symptoms, immediacy of effects and so on.

Augmenting Diagnostic Theory

Much of this implies that at least one other category of outcomes be adjoined to the ones which Sirois proposed: Unknown. That is, one admits the possibility that evidence, for either medical or psychological diagnoses, is simply too puny. This category may be abused in that a manufacturer, employee, or physician may take refuge in a declaration of ignorance. But I suspect that the pressure to diagnose the malady as something with a name is sufficiently high to make even legitimate declaration of ignorance difficult. Lawsuits doubtless increase that pressure. The point is that theory of diagnosis should include criteria for determining when data are not sufficient. Rhetoric to sustain that judgment and to make it palatable to a public which is often unwilling to take "I don't know" for an answer needs to be articulated.

As a practical matter, formal recognition of dissenting opinion on the sufficiency of evidence and the accuracy of a diagnosis is justified. That is, union, management, or other interest groups should be invited to comment on the adequacy of evidence and the commentary published as part of the main report. The practice does not guarantee diagnostic accuracy, but it may help to make competing explanations explicit.

Corruption of Diagnosis

Judging from the remarks of other authors, the diagnosis of contagious hysteria or mass psychogenic illness (MPI) is in danger of corruption. That is, the diagnosis provides a way of shifting the blame for illness from the institution, which may produce hysteria-inducing stresses, to the individual. It is in more danger of becoming a label for a malady in which origins are simply unknown. That is, the real chemical or pathogenic causes of symptoms remain undiscovered despite our efforts and the exclusionary portion of the diagnosis. Or, the diagnosis may become stylish, as in neuromyasthemia, the need for hysterectomy, and the like. The problem of course is not confined to this arena of research. Inflation of body counts in war (Knightly, 1975), of police arrest and conviction rates in new administrations (Campbell, 1975) and the like appear frequently. If Campbell's theory of corruption of social indicators is at all accurate, we ought to expect the inflation of diagnoses of MPI to occur when the social system adopts the diagnosis for administrative purposes.

General Theory

Few methodologists are well informed enough about mass psychogenic illness to dictate changes in the way the theoretician views the problem, and I am not among those few. Nonetheless, it would be unseemly to ignore theoretical features of work on the topic.

Gender Differences. The susceptibility of women to contagious hysteria seems to be over-advertised. Moreover, it conflicts with evidence which, though admittedly scarce, is provocative. Contagious hysteria in wartime, alluded to by Sirois (1974), fainting in blood donation stations and during military vaccination, some epidemic suicide, some outbreaks of violence, and the like, appear to the lay observer to fall legitimately into the category of mass psychogenic illness and *do* involve men. Sympathetic reactions to pregnancy which generate pregnancy symptoms among husbands and interns are at least superficially relevant. The lay observer may be wrong in interpreting this as contagious hysteria and concluding that risk is notable if not high for men. But I've seen no interesting attempt to show why the interpretation is incorrect. It is doubtful that by ignoring gender similarities we will be in any better position to understand what may be more fundamental differences between victims and nonvictims.

Analogues and Relations. Investigators of contagious hysteria often cite the after-dinner yawn as a commonplace example of contagious behavior. One yawn appears to trigger others. But it is not clear that the mechanisms involved in such cases is identical or even similar to one that results in mass psychogenic illness. It does seem to me that, on the evidence of appearance alone, extensions ought to be explored. There are, for instance, rather closer analogues. The evidence that a single individual fainting in a blood donation station triggers other faintings ought to be consolidated and examined for its relation with theory in this area. The rather more upsetting triggers for contagious or at least similar behavior are no less interesting. The vomit produced by one individual in a group in turn is likely to produce nausea among at least a few other members and may produce other symptoms as well. Blood, a sharp cry, a thump on the head, and so on may also produce sympathetic reactions. The analogues suggest a rather larger framework of behavioral phenomena than is typically considered in research on contagious hysteria. More importantly, at least a few are experimentally manipulatable (see Grosser, Polansky, & Lippitt, 1951, for example). Pennebaker's (1979) reports are particularly interesting in this respect. It *is* remarkable to know that perception of temperature, coughing, scratching, sneezing, and so on can be produced among some by merely coughing oneself, or by merely talking about an itch. And both reproducability of the phenomena and its relation to more violent behavior are important to understand. I would suppose that Pennebaker's experiments can be extended to explore a large response surface, using both social and chemical triggers in ways which are not dissimilar from research strategies in marketing, product testing, and chemical engineering. I suppose also that field experiments such as those he has described can help to illuminate currently debatable topics, such as the susceptibility of women, the role of isolation and friendship networks, and the temporal sequence of events.

Synthesis and Quality of Evidence. It is obvious, once said, that theory-oriented synthesis of research on contagious hysteria should take the quality of evidence into account. Sirois (1974) and Markush (1973) recognize the principle, and that is admirable. But the Sirois review and others are less satisfying in that quality of evidence is defined in practice solely by accessibility of information. Setting aside cases in which information is of low quality makes generalization more difficult. But there is lots of precedent for educing misleading theory from faulty data. This includes medical research, where poor experimental design has led to grave errors in judgment about the effectiveness of surgical care (Barnes,

1977), in economics where poor control over quality of information leads to flaws in policy (Morgenstern, 1963), and so on.

Taxonomy. Sirois (1974) appears to be among the few to have tried classifying processes of epidemic hysteria. The classes proposed, based on his review, include: explosive; explosive, preceded by isolated cases; cumulative (slow chain reaction); rebound (massive episode preceded by a small one); massive and diffuse. The taxonomy is useful in simple description, i.e., identifying the forms an outbreak may take. But in the form offered by Sirois, it does not do much more. It is not unreasonable to consider extending it to accommodate variations in predictability, e.g., massive diffuse outbreaks are quite rare, as I understand, whereas other forms, notably explosive preceded by isolated cases, seems to be more common. Nor is it unreasonable to create competing taxonomies that may be more useful in organization (e.g., identification by symptom type or severity), or in prediction, or in explanation.

The Sirois taxonomy does have the merit of forcing one's attention away from the explosive outbreaks and toward other types. It is not clear, however, that it can be linked readily to other taxonomies created to understand hysteria at the individual level, e.g., Laughlin's (Kutash, 1965). Establishing such linkages is worth considering if only to simplify the way we look at the phenomena.

Diagnosis, Theory, and Measurement. Given the difficulty of acquiring information about symptoms during an outbreak, it is no wonder that debate over usefulness of symptoms as an indicator of the nature of illness is vigorous. But even accurate well-articulated information will add only slightly to what one may understand. That is, we will not achieve much by relying on symptoms alone.

Poking through Sax (1979) for example, one finds that carbon disulphide produces nausea, vomiting, headache, depression or irritability, insomnia and chronic fatigue. The same source reports that lead poisoning, of the alimentary form, also produces nausea, vomiting, headache, dizziness, and insomnia frequently. The National Institute of Mental Health (1978) suggests that earthquakes and other natural disasters also produce similar reactions among some people. And there are lots of reports of contagious hysteria, viral infections, etc. producing the same reactions.

Sax (1979) warns against imputing the origins of the problem from symptoms alone for the case of dangerous industrial materials. There is no good reason to ignore that advice for illness with psychogenic origins. This leads us to theory and what information one needs to discriminate between psychogenic and other illness. To be sure, behavioral symptoms are important. Some origins can be ruled out because their effects are more specific, more dramatic, or different, a matter of contemporary toxicology as I understand it.

Some theory, explored by Kerckhoff et al. (1965) among others, suggests that contagion operates through friendship networks, at least at later stages. To the extent that it does so regularly, the network itself becomes a useful element in diagnosis. And the theory certainly implies that information about networks ought to be obtained. That networks may be confounded with other characteristics, such as seating arrangement, is also clear. This implies obtaining information about physical proximity and traffic patterns as well.

At least in principle, inducement of relapse, discussed earlier, is a useful device for discriminating between illness with psychogenic origins and illness caused by other problems. That is, if in the absence of toxic environmental influence, episodes occur with visits of investigators, supervisors, or reproduction of earlier conditions, it seems to me that the case for psychogenic illness can be made more persuasively.

The benign morbidity of contagious hysteria is, according to Sirois (1974) at least, a useful discriminator. It is consistent with his theory, and this suggests that follow-up studies are warranted to assure that effects are indeed insignificant. That theory, however, is not entirely consistent in that it ignores rather more serious forms of what some would regard as contagious hysteria, e.g., suicides which occur in clusters.

For Kutash (1965), some of the criteria to be used in distinguishing between hysterical and organic disturbances must include the ''*belle indifference* of the hysteric in relation to his symptom, the frequent failure of the disfunction to follow a correct anatomical or mere distribution pattern, contradictions such as the person who is hysterically blind but avoids bumping into persons or objects, and the susceptibility of the symptoms to removal by hypnosis or suggestion [p. 964].'' This standard has some promise when the purported effects of toxic agents are well-understood and the correct anatomic patterns known. Judging from the number of new chemicals whose effects are not understood, however, the standard is useful only for a small class of dysfunctions. An illustration of the use of suggestion which appears to be available is the data computation center case: weather inversion was labeled as the origin of illness (Stahl & Lebedun, 1974). The ethical implications of constructing an explanation which is patently false have not been worked out. More to the point, resorting to suggestion or recording the consequence of accidental suggestion is warranted, since it too may enhance identifiability of the illness.

At least some theory based on laboratory study as well as the field tests holds that individuals who are susceptible to hysteria score high on measures of neuroticism and at the intermediate level on measures of extroversion (Eysenck & Beech, 1971). Assuming that the evidence is adequate, the implication is that indicators of neuroticism and extroversion be used in investigations to verify the illness. Exploiting this approach is likely to be difficult with existing indicators; development of less obtrusive indicators may be possible, but I do not know the area well enough to predict accurately. Related theory also stresses the use of various forms of therapy and tests of therapy for chronic cases (Eysenck & Beech, 1971). Again, the effectiveness of the therapy may help to establish that the label attached to an illness is proper. But follow-up study of contagious hysteria is rare, and opportunities to employ therapies may be still more rare.

Physical contact or visual observation of an episode is purported to be a reasonable device for helping to discriminate between psychogenic and other illness, and is consistent with some theory of contagious hysteria (see Chew et al., 1976, for instance). It is rather more complicated an indicator, of course, in that viruses and the like are as plausible a cause of illness in these instances; visual contact may be a surrogate indicator of proximity to a gas or chemical producing the result. To the extent that these can be ruled out, physical and visual observation are candidates for assisting discrimination and so ought to be recorded.

Finally, though stress is said to be a major influence on the occurrence of mass psychogenic illness, it is not yet clear how useful a discriminator it is. That is, many environments are stressful, but no outbreaks occur. Perhaps they do occur and we are not informed. It may also be argued that episodes occur in environments that are not stressful. Because no one has any information about these at all, the argument cannot be resolved. All this implies greater attention to surveys as a vehicle for understanding how to characterize the problem. We are not entitled to blame stress, except tentatively, unless there are data that show that outbreaks are more likely in high stress settings than low stress settings. Even then the stress adds incrementally to prediction. And stress itself is as likely to be a symptom as well as cause of an outbreak.

Though a good deal is said about them, a few other indicators appear to be less useful than advertised, at least in theory. This includes the purported role of sex: Women, it is said, are more susceptible. Yet counterevidence, if not strong, is at least plausible (see earlier remarks). Similarly, education level and socioeconomic class are said to assist in discrimination. But we have illustrations of physicians and nurses being affected. And, to my knowledge, there has been no concerted attempt to examine, for example, middle managers of companies to assay frequency of the event.

AMELIORATIVE ACTION AND ITS EFFECT

Over the last 300 years or so, the treatments invented for victims of contagious hysteria have had the merit of being entertaining if not always effective. Soldiers were ordered to succor a convent's mewing nuns during the 18th century by beating them with rods supplied, evidently, by the local sheriff (Markush, 1973). Individuals so equipped and so disposed organized themselves into a religious order to assist the convulsionnaires of St. Medard during the same period. Drowning, torture, hanging, burning, dismemberment, and similar remedies have each had their time in the limelight. Even if the remedies eliminated both victim and disease, there were elaborate efforts to describe treatment and at least a few attempts to rationalize why they worked (Kors & Peters, 1972). It is not unreasonable to expect similarly conscientious, if less lurid, attention to detail in reporting about contemporary treatment of mass psychogenic illness.

Reporting

But the specific actions taken to remedy a problem and the rationale for use in the particular episode, irrespective of their probable effectiveness, are not often well reported. Sedatives are reported to have been used, but there is no specification about type, time, dosage level and so on; no mention is made of placebo level doses. Isolation is employed but no information about timing, duration, and level of isolation is given. Some victims are hospitalized but we know little about duration of hospitalization or therapies provided, and the basis for hospitalization. In the absence of more specific information, it is very difficult to justify a major investment in understanding whether a particular action works to reduce the problem and theory to understand why an action works is gratuitous.

Ameliorative Action and Delivery of Treatment

One of the few reliable rules of social research is that treatments are rarely delivered as advertised. Ameliorative action may differ systematically from plan, as when many individuals fail to take a prescribed sedative because of memory failure, distrust, or superstition. Or the deviation may be asystematic, as when individuals are persuaded to varying degrees that their problem is psychological rather than chemical in origin. The phenomena is, of course, not confined to attempts to cope with epidemic hysteria, though it may be especially difficult to assay in these cases. It occurs with fair regularity in medicine, clinical psychology, welfare programs, and education (Boruch & Gomez, 1977).

The rule implies that at least crude verification of treatment is necessary. Independent testimony about who took what action, how soon, how often, and for how long is reasonable in principle, if difficult to practice. But there is little recognition of the rule in reports on mass psychogenic illness.

The Response Expected Under Treatment

The effects generally anticipated from treatment are usually specified in reports. One would like to see the individual calmed as soon as possible, the contagion minimized, a return to work, and minimum disruption following return to work.

But it is difficult to find more than temporary attention to these expectations. The absence of longer term studies will lead us inevitably into scientific sins if history is any standard. Barnes (1977), for example, argues that the popularity of useless surgical treatments for ''displaced'' organs, for constipation, and for control of ''kenetic drive'' during the early 19th century is due partly to the absence of follow-up research. It is unlikely that investigations of the effectiveness of action taken to ameliorate psychogenic illness will fare better if such studies are not executed.

The effects which are *not* anticipated are no less important especially if they are negative. Sirois (1974) seems to be among the few investigators to recognize this in a general way, by suggesting that some approaches to treatment, such as exorcism, may exacerbate the problem. But the negative effects of treatment are not well documented for clinical therapy in general much less for the care of hysterics. Experts such as Mowrer (1965) issue periodic warnings about the iatric inflation of personality disorders as a function of treatment, but experimental research is sparse (see for example Lieberman, Yalom, & Miles, 1973, on encounter groups for an exception). Admitting the possibility of negative effects of what is ostensibly benign treatment is important. For if we don't look for these, we won't find them, and we will have deceived ourselves unnecessarily.

The Positive Effects of Remedial Action

Reports of mass psychogenic illness often contain the authors' judgments about the effectiveness of action taken to remedy the problem. The grounds for judgment are not always very clear, however.

Chew's (1978) report, for instance, is confusing in that the effects of chlorpromazine on victims are first said to have been negligible. But the author goes on to report that after the girls had calmed down, they were sent home, implying that chlorpromazine could have been effective. Knight et al. (1965) say that many students continued to have attacks in spite of taking the prescribed medication on a regular basis, but they provide no detail on how regularity was monitored or on how many students in the medicated group had attacks. Public confrontation and exhortation are, according to Sirois (1974), "usually useless and aggravate the outbreak [p. 19]." That is plausible, but he provides no evidence. On the other hand, Chew (1978) maintains that dousing a victim with water discouraged fakery and implies that it may have impeded further attacks. Conversation itself is said to be helpful, e.g., Colligan and Stockton (1978) emphasize the psychologists' "making clear to the individual that symptoms are real," and that too is plausible but the evidence is informal. Counterfeit explanation led to cessation of attacks in the data computation center case (Stahl & Lebedun, 1974), but we cannot tell from the report whether and for how long individuals were monitored after the first attack.

The point is that it is difficult to decide what works based on such evidence, unless one is willing to rely entirely on the investigator's expertise. Even with very good evidence, decisions about whether an action ameliorates a problem is not easy. Gaps in reporting exacerbate the problem.

It does make sense to arrange the evidence on the impact of an action in order of persuasiveness. Consider, for example, treatment of the victim. To the extent that the effect of action is immediate and large, and there is at least crude theory to link action and response, and there are no other uncontrolled influences on the victim, attributing the result to action is not unreasonable. In the easiest cases, the effect of sedation will be almost immediate and consistent with theory of how sedation works. And the victim will not have responded to placebo. Variation in individual response will complicate matters of course, but conformity to the stereotype ought to be recognized. To the extent that response to remedial action is delayed, taking credit for control of the problem is risky: People become calm even without sedatives. To the extent that the observed effect is small, taking credit is riskier still, because small variations in human behavior are the rule. To the extent that there are a great many actions taken: sedation, isolation, conversation, etc., attributing positive response to one is not likely to be warranted. The most ambiguous evidence—delayed, small response, etc.—ought to be identified as such.

Most remedial action is cheap and so there is little justification for attempting, in small formal experiments, to isolate their relative effectiveness. There is, however, some theoretical justification. It is important, for example, to understand whether a placebo works relative to drugs in sedation. And it does seem possible to design small tests if victims are treated in isolation, and to recommend systematic introduction of treatment tests rather than *ad hoc* action as part of instruction to physicians who may have to cope with these problems. It also seems conscionable to manipulate leave. That is, assigning a victim to a week of home rest is ethical, but may encourage other attacks as home rest is a reward of sorts. I suppose that it can be argued that requiring very early return to work, rather than home rest, is ethical too, in that getting victims back on their feet is important. And the implication is that one may be justified in running controlled trials of the alternatives in this arena as in other arenas when effectiveness of treatment is debatable. Though the design of small studies of the sort is worth considering, rather more thinking needs to be done on the topic. Without design, and in the absence of immediate large effects of treatment, it is doubtful that we will learn very much about the sense in which treatments work or produce negative effects for the individual victim.

Some investigators also allude to the on-site objective of minimizing or reducing the number of attacks. The objective makes little sense in that we usually do not know what level of contagion will appear in the absence of extraordinary action. Exceptions *might* include the Singapore work where reappearance of mass psychogenic illness is, judging from Phoon's remarks at this conference, not uncommon. It may then be reasonable to base one's judgment on peak levels of psychogenic illness encountered in the past. Standards weaker than this weak one are also possible, e.g., the limits implied by friendship networks in which attacks spread. But I can think of no stronger bases for evaluating statements about how attacks were prevented.

External Therapy and Comparative Tests

Despite the large body of literature on individual hysteria and the presumption that a contagious variety underlies mass psychogenic illness, there does not appear to have been much research on treatment of victims following an outbreak. That is, no formal tests of desensitization therapy or other strategies have been undertaken to illuminate the problem. It is not clear that available therapy would be useful: That is the reason for suggesting comparative tests. Nor is it yet clear that a testing program is justified on economic grounds. Assuming that it is, those tests might be undertaken in the laboratory, the therapeutic setting, or on site. Whatever the location, if some therapies are found to work, they may provide clues about more effective action during an outbreak, or at least some indirect evidence that action is likely to have been helpful rather than harmful. They may have the benefit of increasing the individual's resistance to influences that provoked the outbreak in the first instance. And they will almost certainly help to understand the etiology of psychogenic illness.

ANALYZING COUNT DATA

The Effect of Nonresponse on Count Data

The unsettled nature of an episode of psychogenic illness is almost certain to produce missing data. For example, despite the Colligan et al. (in press) plan to sample victims and nonvictims in the electronics plant outbreak, they succeeded in obtaining information from only 81 of the 150 individuals to whom questionnaires had been given. They deserve credit for the attempt to sample and report results, especially because such reports are infrequent. The response rate, however, does suggest that that analysis of the inferential biases engendered by missing data is warranted.

To illustrate what might be done in simple sensitivity analysis, consider a situation in which response rate is imperfect but the obtained sample is large and responses are accurate. Suppose that only 54% of the target group responds, as in the Colligan et al. case. Their report suggests that 58% of the respondents had headaches during the outbreak. The suggestion is accurate (ignoring sampling error) if the people who respond are a random sample of the target sample. But nonrespondents may differ from respondents in education level, compulsivity, and other ways, as they often do in other action research. One may construct a less presumptive description of headache rate by recognizing nonresponse and computing arithmetically permissible upper and lower bounds on headache rate in the target sample of 150. Doing so suggests that at minimum, 31% had headaches and at maximum, 78% did so.

Developing crude bounds on the possible value of a proportion for a particular sample or developing more general tables requires only simple arithmetic. Table 6.1 has been constructed to facilitate review of proportions based on imperfect response rate. A response rate of 50%, given in the first column of the table, and an observed proportion of 75%, given in the second column, yields a permissible arithmetic range of 38–88%. That is, between 38% and 88% of the target sample have the trait in question.

TABLE 6.1

Arithmetic range for the true proportion of individuals with a given trait when the overall response rate is 30%, 40%, . . . 100%, and the proportion of observed Yes responses to inquiry is 0%, 25%, . . . 100%.

Overall Response Rate	*Observed Proportion of Yes Responses*	*Minimum True Proportion of Yes Responses*	*Maximum True Proportion of Yes Responses*
100%	50%	50%	50%
80%	100%	80%	100%
	75	60	80
	50	40	60
	25	20	40
	0	0	20
70%	100%	70%	100%
	75	53	83
	50	35	65
	25	18	48
	0	0	30
60%	100%	60%	100%
	75	45	85
	50	30	70
	25	15	55
	0	0	40
50%	100%	50%	100%
	75	38	88
	50	25	75
	25	13	63
	0	0	50
40%	100%	40%	100%
	75	30	90
	50	20	80
	25	10	70
	0	0	60
30%	100%	30%	100%
	75	23	93
	50	15	85
	25	8	78
	0	0	70

The table assumes perfectly valid response and ignores sampling variation.

I have used the word arithmetic here because although such tables are useful they ignore statistics. Indeed, one ought to construct little confidence intervals, if the target sample and actual sample is selected from some larger target population. Sample-to-sample variation ought to be recognized in Colligan et al. (in press) as well as in a sensitivity analysis table of this sort. One can further presume that responses will be imperfect: A respondent may affirm falsely that he has a symptom, or falsely deny that he has it. If the rate of false positives and false negatives can be guessed or estimated from side studies, then more elaborate tables can built to help assay credibility of original data.

The task of comparing groups is similarly complicated by nonresponse, especially if nonresponse rate differs from one group to the next, or from one time to the next. Suppose, for example, that one is fortunate enough to have a headache count before an outbreak. Assume that response rate is 50% and the observed proportion of people in this sample who admit to headaches on a given day is 25%. After an outbreak one might obtain *exactly* the same response rate and headache rate. Yet it is possible for headache rate to have changed from 15% (the minimum) to 63% (the maximum), the change being masked by poor response rate. It is similarly possible for a group of victims and a group of nonvictims to have exactly the same response rate of 50% and rate of affirming headaches, 25%, yet actual rate of headache can differ notably, as much as 50%.

As one might suppose, the arithmetic range becomes larger with higher level of nonresponse, lower validity in response, and smaller sample. There is no substitute for good data. But the ambiguity might be reduced if stereotypical forms of bias are known. One may find, for example, that respondents and nonrespondents in a particular case do *not* differ from one another in education level or other characteristics. The information makes it more reasonable for some analysts to assume that the headache rate found in the respondent group would also be found in the nonrespondent group, for instance. And it reduces one's dependence on computation of the sort described. That matching on the basis of only partly relevant variables, such as education, may itself produce bias in analysis is also clear. For stereotypical forms of this bias, see text and references to work in chemistry, education, and medicine in Campbell and Boruch (1975).

Relative Counts

Although the number of people affected in an institutional episode are usually reported, the number is not always informative. Meaningfulness of reported incidence depends partly on the number of individuals who might be affected.

That is, the number of victims relative to the total number in a group more or less at risk is important but rarely reported. That 25 individuals out of 25 were affected seems to imply something different from a case in which 25 out of 100 or 1000 are affected. In particular, I should think that physically dangerous environmental influences are more likely in the first case, and psychogenic illness, or environmental insult, or both are more likely in the second. Further, theories offered to explain hysterical contagion may hinge on isolation, and on the proportion of a group who are friends, e.g., Kerckhoff et al. (1965). In any case, a report of the absolute number affected is misleading, constrained as it is by physical characteristics of the environment and the size of the administrative unit. A ratio helps direct attention to the unaffected group, and that is warranted scientifically. Finally, we ought to know whether the relation is reasonably stable from one episode to the next, for this will help us to predict crudely at least, and may help us to understand better the effects of remedial action.

How base rates of the sort ought to be computed is not at all clear, however. Specifically, how should we define the total group, the denominator of the computation? Should we use the number of individuals within an administrative unit in which an epidemic occurs? Or, should we use the number of individuals who occupy the same room or floor, or who are otherwise in visual contact or physical proximity? Because any convention is likely to be premature, and because computation of each ratio is cheap, providing ratios based on each is probably warranted.

ARCHIVES AND SECONDARY ANALYSIS

The idea that one ought to store data generated by the investigation of mass psychogenic illness, and to make it available for independent secondary analysis, is implicit in Colligan and Stockton's (1978) work. It is consistent with a scientific tradition that encourages skeptical review, and with a more recent tradition of regarding expensive data as a national resource. It is partly a consequence of this tradition that journals such as *Anesthesiology,* for example, occasionally publish the entire data set used as a basis for research along with the pertinent prose. Others, such as the *Journal of Personality and Social Psychology,* maintain an editorial policy that requires that authors make raw data available for reanalysis. Formal publication of second opinion, and especially of dissenting views, along with original analysis is similarly consistent with a principle of ready scientific access to information, e.g., the U.S. General Accounting Office's publication of criticism of a GAO investigation. The same practices could be adopted to improve quality of reporting and access to data on psychogenic illness.

Publication of all pertinent material is not justified, much less feasible. Yet logs, letters, memos, and microdata may be fundamental to competing analyses of a given data set by analysts outside the original group of investigators. An archive can help to the extent that it provides flexible access to all relevant material. It may also be justified on grounds other than facilitating secondary analysis of particular cases. An archive can help to inform one's decision to pool data across sites and can facilitate the pooling process in the interest of better statistical power. Protocols for investigation will help to increase uniformity of reporting, and an archive is a reasonable vehicle for monitoring adherence to protocols, for integrating material, and for improving the quality and flexibility of protocols. In any event, an archive makes good economic sense in making expensive data work repeatedly and at low cost in this as in other scientific areas (Boruch, Wortman and Cordray, 1981; Hedrick, Boruch, Ross, 1978).

More generally, the notion of statistical archives designed for such purposes is consistent with federal statistical policy (President's Reorganization Project, 1978) and law (Privacy Protection Study Commission, 1977). It may be possible to expand existing archives, in the interest of economy, as Dollar and Armbacher (1978) suggest in their description of the National Archives' machine-readable data program. The existence of pertinent professional organization, such as the International Federation of Data Organizations, makes world-wide cooperation possible.

ACKNOWLEDGMENT

Background research on methods has been supported by the National Science Foundation (DAR 7820374). James Pennebaker, Mike Colligan, and Lisa Simon kindly provided editorial advice on an earlier draft of this paper. Colligan was especially helpful in supplying leads to reports on the topic.

REFERENCES

Asato, L. *Mass contagion*. Student Paper. Psychology Department University of Southern California, 1969.

Assakul, K., & Proctor, C. H. Testing independence in two-way contingency tables with data subject to misclassification. *Psychometrika*, 1967, *32*, 67–76.

Barnes, B. A. Discarded operations: Surgical innovation by trial and error. In J. P. Bunker, B. A. Barnes, & F. Mosteller (Eds.), *Costs, risks, and benefits, of surgery*. New York: Oxford, 1977.

Beals, R. L. *Politics of social research*. Chicago, Aldine, 1969.

Boruch, R. F., & Gomez, H. Sensitivity, bias, and theory in impact evaluations. *Professional Psychology*, 1977, *8*, 422–434.

Boruch, R. F., & Cecil, J. S. *Assuring the confidentiality of social research data*. Philadelphia: University of Pennsylvania, 1979.

Boruch, R. F., Wortman, P. M., & Cordray, D. S. (Eds.) Reanalyzing program evaluations. San Francisco: Jossey-Bass, 1981.

Campbell, D. T. Assessing the impact of planned social change. In G. M. Lyons (Ed.), *Social research and public policies*. Hanover, New Hampshire: University of New Hampshire, 1975.

Campbell, D. T., & Boruch, R. F. Making the case for randomized assignment to treatment by considering the alternatives. In C. A. Bennett & A. A. Lumsdaine (Eds.), *Central issues in social program evaluation*. New York: Academic, 1975.

Chew, P. K. How to handle hysterical workers. *Occupational Health and Safety*, March/April, 1978. 50–53.

Chew, P. K., Phoon, W. H., & Mae-Lim, H. A. Epidemic hysteria among some factory workers in Singapore. *Singapore Medical Journal*, 1976, *17*, 10–15.

Cochran, W. G. Errors of measurement in statistics. *Technometrics,* 1968, *10,* 637-665.

Cohen, B., G. F., Colligan, M. J., Wester, W., & Smith, M. J. An investigation of job satisfaction factors in an incident of mass psychogenic illness at the workplace. *Occupational Health Nursing,* 1978, *126,* 10-16.

Colligan, M. J., & Stockton, W. The mystery of assembly-line hysteria. *Psychology Today,* June 1978, 93-116.

Colligan, M. J., & Murphy, L. R. Mass psychogenic illness in organizations: An overview. *Journal of Occupational Psychology,* 1979. *52,* 77-90.

Colligan, M. J., Urtes, M. A., Wisseman, C., Rosensteel, R. E., & Anania, T. L. An investigation of apparent mass psychogenic illness in an electronics plant. *Journal of Behavioral Medicine,* 1979, *5*(3), 297-309.

Cronbach, L. J., Gleser, G. C., Nanda, H., & Rajaratham, N. The *Dependability of Behavioral Measurements.* New York: Wiley, 1972.

Dollar, C. M., & Armbacher, B. I. The National Archives and secondary analysis. *New Directions in Program Evaluation,* 1978, *4,* 1-6.

Ebrahim, G. J. Mass hysteria in school children: Notes on three outbreaks in East Africa. *Clinical Pediatrics,* 1968, *7,* 437-438.

Eysenck, H. J.,& Beech, R. Counterconditioning and related methods. In: A. E. Bergin and S. L. Garfield (Eds.) *Handbook of psychotherapy and behavior change: An empirical analysis.* New York: Wiley, 1971.

Grosser, D., Polansky, N., & Lippitt, R. A laboratory study of behavioral contagion. *Human Relations,* 1951, *4,* 115-142.

Hedrick, T. E., Boruch, R. F., & Ross, J. 1978. On insuring the availability of evaluative data for secondary analysis. *Policy Sciences,* 1978, *9,* 259-280.

Herrick, R. The historical perspective. In N. I. Sax (Ed.), *Dangerous properties of industrial materials* (Fifth Edition). New York: Van Nostrand Reinhold, 1979.

Kerckhoff, A. C., Back, K. W., & Miller, N. Sociometric patterns in hysterical contagion. *Sociometry,* 1965, *28,* 2-15.

Knight, J. A., Friedman, T. I., & Sulianti, J. 1965. Epidemic hysteria: A field study. *American Journal of Public Health,* 1965, *55,* 858-865.

Knightly, P. *The first casualty.* New York: Harcourt Brace Jovanovich, 1975.

Kolata, G. B. Is labile hypertension a myth? *Science,* 1979, *204,* 489.

Kors, A. C., & Peters, E. *Witchcraft in Europe, 1500-1700: A documentary history.* Philadelphia: University of Pennsylvania Press, 1972.

Kutash, S. B. Psychoneuroses. In B. B. Wolman, (Ed.), *Handbook, of clinical psychology.* New York: McGraw-Hill, 1965, 948-975.

Lieberman, M. A., Yalom, I. D., & Miles, M. B. *Encounter groups: First facts.* New York: Basic, 1973.

Lord, F. M., & Novick, M. R. *Statistical theories of mental test scores.* Reading, Mass: Addison-Wesley, 1968.

Markush, R. E. Mental epidemics: A review of the old to prepare for the new. *Public Health Reviews,* 1973, *2,* 353-441.

Mausner, J. S., & Gezon, H. M. Report on a phantom epidemic of gonorrhea. *American Journal of Epidemiology,* 1967, *85,* 320-331.

McEvedy, C. P., & Beard, A. W. A controlled follow-up of cases involved in an epidemic of benign myalegic encephalomyelitis. *British Journal of Psychiatry,* 1973, *122,* 141-150.

Morgenstern, O. *On the accuracy of economic observations.* Princeton: Princeton University Press, 1963.

Mowrer, O. H. Learning theory and behavior therapy. In B. B. Wolman (Ed.), *Handbook of clinical psychology.* New York: McGraw-Hill, 1965.

Murray, J. R. *Statistical models for qualitative data with classification errors*. Ph.D. Dissertation, University of Chicago, Chicago, 1971.

National Institute of Mental Health. *Field manual for human service workers in major disasters*. Washington, D.C.: U.S. Department of Health, Education, and Welfare, 1978.

Pennebaker, J. *Social and perceptual factors affecting symptom reporting and hysterical contagion*. Paper presented at the Symposium on the Diagnosis and Amelioration of Mass Psychogenic Illness, May 30–June 1, 1979, Chicago.

President's Reorganization Project. *Issues and Options*. Washington, D.C.: U.S. Office of Management and Budget, 1978.

Privacy Protection Study Commission. *Personal privacy in an information society*. Washington, D.C. U.S.: Government Printing Office, 1977.

Riecken, H. W., & Boruch, R. F. (Eds.), *Social experimentation: A method for planning and evaluating social programs*. New York: Academic Press, 1974.

Sax, N. I. *Dangerous properties of industrial materials*. New York: Van Nostrand Reinhold, 1979.

Sirois, F. Epidemic hysteria. *Acta Psychiatrica Scandinavica. Supplementum*. 1974.

Stahl, S. M., & Lebedun, M. Mystery gas: An analysis of mass hysteria. *Journal of Health and Social Behavior*, 1974, *15*, 44–50.

7 Yes Virginia, There Really is a Mass Psychogenic Illness

Jerome E. Singer
Uniformed Services School of Medicine

INTRODUCTION

Scientific investigation has never been value free. The scientific equivalent of apple pie and motherhood are truth, objectivity, and skepticism; it is an article of faith that these basic beliefs are always in operation. At another set of levels, political considerations bring values to science. In addition, scientists live at particular times in particular cultures and bring society's values into the laboratory with them. Even the funding of research brings values to an investigation. In an era when science is too expensive to be conducted by self-support, the research agendas and priorities of foundations and funding agencies deeply influence what is valued and what gets done. Most of these value systems are examined by scholars who study science as a social institution; the day-to-day worker does not usually confront these issues.

The further scientists are removed from the applications and implications of their work, the less likely they are to attend to issues of value and political import. Conversely, the closer they are to the policy and social consequences of their science, the more salient these issues will be. The study of Mass Psychogenic Illness (MPI) is a crucible where scientists from academia, relatively sheltered from the hurly-burly of politics and policy, are brought together with others who are directly involved in policy decisions.

For the academic psychologist or sociologist, MPI is a fascinating topic. It is a real world example of a variety of phenomena, ranging from social contagion, social influence theories, normative structures, attribution theories, to whatever else is currently fashionable in the behavioral and social sciences. For the scientist in the workplace, however, the occurrence of an epidemic of unexplained physical symptoms represents a value-laden challenge. If an immediate cause, such as a gas leak or volatile chemical, can be identified, the political questions of what remedial actions at what costs at what risks can be openly addressed. The task is much more difficult when no ostensible reason can be found for a dramatic outbreak of symptomatology. If the victims show no physical deficit upon medical examination, the case is even more murky.

Scientists of various opinions can differ as to the cause of the outbreak with a continuum of possibilities ranging from a purely social etiology at one end to a purely physical etiology (as yet undiscovered) at the other. There are strong incentives for scientists with union or consumer activist affiliations to prefer interpretations favoring physical causes. Those involved with management or industry prefer interpretations favoring social or psychological causes. The reasons for these preferences are clear. If the causes are physical, then workers are entitled to a variety of benefits—such as compensation and medical care; governmental agencies are empowered to act to alter working conditions, the workers are not labeled "unstable" or "hysterical," and the conditions in the workplace become a potent bargaining issue. Most importantly, the attribution to a physical cause encourages improvements in the industrial hygiene of the workplace to protect workers against the effects of a toxin that is discovered only belatedly. If the causes are psychosocial, safety records are maintained, environmental enforcement actions are precluded, insurance premiums are not escalated, expensive additional chemical monitoring is avoided, health care and compensation costs are minimized, and productivity criteria need not be modified.

These issues have been joined. Whenever an outbreak of physical symptoms occurs, scientists and hygienists from labor, industry, and government are all involved—if not in the direct investigation, then at least in the interpretation and implication of the findings. And, just as all parties have vested interest in the resolution of the etiology of every specific incident, they have similar interests in the general study of MPI. Most of the academic papers in this volume have been produced in an atmosphere far removed from the workplace, but they are liable to critical examination from groups of readers who regard the stakes as immediate and important.

Among the people who deal with industrial sickness, hygienic matters, work disturbances and the like, there seems to exist a group for whom the notion of psychogenic illness still seems a bit fishy, a possibly fictional notion. These people, who are not quite sure whether MPI exists or not, raise in one form or another the general question: Is there really anything called mass psychogenic illness that is other than a residual category? That is, you find a group of people getting sick, you investigate, you measure everything you can measure, you crank up the chromatographs, and when you still can't find any physical reason, you say "well, there's nothing else here, so let's call it a case of MPI." And, thus, does another hypothetical event become reified by word magic. Although this criticism has a surface appeal, we ought to explore the fact that a case can be made that mass psychogenic illness is a genuine phenomenon. We first have to distinguish between the individual case and the general one—that is, "Is this a particular instance of MPI?" rather than "Does MPI occur at all?" Now, there is always the possibility that any individual case has an alternative explanation. The alternative explanation will usually be that some noxious agent that you have failed to discover is the presumptive cause of the outbreak of illness. However, when we accumulate a large enough number of cases and we can show systematic consistency across them, as several of our review papers have pointed out, it becomes less and less plausible that each one of these similar cases has had one or another alternative etiology or alternative genesis for the outbreaks of illnesses that have occurred among the workers.

Secondly, most of the alternative explanations have been concerned with particular kinds of noxious elements such as pesticides, solvents, or air pollutants. But we shouldn't lose sight of the fact that although many instances are found in industrial settings, other outbreaks of mass psychogenic illness occur outside the workplace. MPI is found in schools, where the probability of solvents and the like are much less likely, and in fact, one early classic case, the Phantom Anesthetist of Mattoon (Johnson, 1945) occurred outdoors in the street. We don't know for sure what small town Illinois air quality was like in 1944, but it is more than reasonable to suppose that it was not anywhere near as bad as the airborne noxious agents encountered in more recent factories.

Finally, we should note that in the cases of mass psychogenic illness that do occur within the workplace, the illness affects selected people. These people are usually ethnically similar, as in the case of the Singapore studies where primarily the Malays were involved, or for outbreaks in the U.S. where the symptomatology appeared only in the sociometrically similar within otherwise homogeneous groups.

It's very difficult to conceive of a physical agent that would so selectively attack people who happen to be friends with each other. There might be a case for the existence of differential genetically-induced susceptibility among different ethnic groups. This is a possibility for some of the Singapore studies, but it hardly helps explain the sociometric differences within the U.S.

TOXIC VS. SOCIAL CAUSES

Mass psychogenic illness versus a toxic cause for a particular outbreak need not be viewed as alternative explanations. It is not an either/or phenomenon. It is not the case that somebody either gets sick from a chemical toxin (water or airborne) or else has mass psychogenic illness. Both factors can be involved. However, the case most often advanced is that for every time mass psychogenic illness is presumed to occur, there may be a physical factor that is yet to be ruled out. The shoe could, in fact, be put on the other foot, and the case could as easily be made that for every time a physically noxious agent is present, the concomittant psychological factors have not been definitely ruled out, and perhaps psychogenic illness is operating in these instances also.

The ways in which things are ruled in or out and the logic of scientific argument makes it impossible to really prove one or the other extreme possibility. In any event, the main theme is, "don't rule out MPI just because a physical agent is present." As a matter of fact, physical and psychogenic factors may potentiate each other. It may very well be that physical agents that are below toxic levels have effects only because an MPI or a similar process are potentiating them: They could very well be operating in a synergistic fashion.

Although it may be impossible to prove, or at least establish to the satisfaction of a prudent nonpartisan observer, the relative contribution of physical and psychogenic factors in triggering an illness epidemic, there is some utility to looking for conditions of plausibility. Let us consider two possible circumstances under which no physical agent is detected but one is presumed responsible for the illness outbreak. The first case is one in which the argument is offered that the offending toxin has not been measured—for example, perhaps only industrial solvents have been monitored, but the culprit is a type of sewer gas. Because it is logically impossible to discount an unidentified toxin, other questions must be asked. (Clearly, caution is called for, as there are numerous instances of strange outbreaks, seemingly psychogenic, for which physical causes were ultimately found.) Are the affected people in physical proximity or scattered about the site? The more scattered they are, the less likely that a local concentration is causing the illness. Are all of the affecteds friends, or ethnically, or socially similar? The more homogeneous the affected group, the less likely that some general physical agent is responsible. Does medical examination yield a coherent set of symptoms or syndromes? The more diverse the examination results, the less likely was the illness to have been caused by physical factors. A series of questions such as these can render the competing explanations more or less plausible.

An easier argument to treat is the one that claims that a physical agent either in too low a concentration to be measured or one that is measurable but present in a concentration below the toxic threshold, has caused a reaction in a number of particularly susceptible, hypersensitive individuals. At the extreme, some commentators have claimed that concentrations on the order of 4 or 5 molecules per cubic meter of air may be sufficient to incapacitate the most sensitive individuals. It may be instructive to examine the implications of that worst case. Physiologists suggest that a good approximation of human respiratory volume is that humans will respire about 6 liters/min (12 breaths/min at 500 ml/breath). To be on the conservative side let us assume a volume of 10 liters/min and grant that entirely different gas mass will be inspired at each breath. It will take then about 100 min for a cubic meter of air to be respired. If we further assume that each molecule of toxin stimulates a receptor, each receptor a neuron, each neuron a nerve fiber, and all other necessary ideal conditions, we can see that with even 20 molecules per meter, the maximum signal rate is one impulse every 5 min. For most independent evaluators, this is probably far below the values needed to seem plausible. Each person can determine for him or herself what concentrations of toxic substances are necessary before an explanation based on hypersensitivity becomes plausible on neurophysiological grounds. It should also be noted that a hypersensitivity explanation must also satisfy the constraints of friendship, lack of physical proximity, ethnic similarity, and the sudden disappearance of the symptoms that characterize so many outbreaks of seemingly unexplained illness.

Most medical phenomena are rarely susceptible to incontrovertible evidence, and informed judgments usually rest on the relative probability of each of a set of competing explanations. The point of the exercise in the preceding paragraph was not to deny the potential existence of very susceptible victims or of extremely potent toxins, but to point out that in cases where no palpable toxin is identified the postulation of an as yet to be identified one may involve a chain of assumptions each of which is of low probability. There is, in fact, little difference in the tangible basis of MPI versus a physical toxin. At the limit both rest on a series of unverifiable assumptions.

THE NORMALITY OF MPI

Another point which needs to be stressed because it seems to underlie some of the skepticism regarding MPI is whether the processes are normal. We regard illness in response to a toxin or something in the air as unhealthy, but perfectly normal; we don't think it's strange that somebody becomes ill when they breathe a substance that they shouldn't be breathing. MPI processes should not be regarded as abnormal either. They are normal human social reactions to the conditions that instigate them. In the same way as a physical sickness to a physical toxin would be considered the normal reaction so should MPI be regarded. It's not healthy, it's not good, and it's certainly not something you go out and do in your spare time; but there is nothing abnormal about it. The particular explanation that I favor for MPI (to be discussed in later chapters) derives from a particular social psychological theory called social comparison processes. And the social comparison processes theory was derived exclusively for a discussion of how people normally function in their everyday situations. The fact that it can fit MPI emphasizes the role that normal processes, such as information processing, evaluation of belief, or comparison with others, have in the etiology and the transmission of MPI.

Now, all of these factors imply that MPI should be attributed not to a particular person, this, that or the other victim, but rather to the conditions and the precipitating factors, things in the work group, some of which may be in the control of the organization or the company, some of which may not. The person who gets sick in an outburst of MPI is a victim, not a dupe nor a loony. It's somebody who is suffering from an illness brought about by conditions that are no less real, even though they don't show up on a chromatograph, than somebody who is sick because of a specific toxin. The argument, whether implicit or overtly made, that a physical toxin implies a culpability but MPI does not, we regard as specious. If management is to be accountable for conditions that produce illness in the workplace, there is logically no difference between physical and social conditions that have the same effect. We recognize that the implications of this position are serious and expensive, involving as they do complicated issues of health benefits, compensation, grievances, etc. The point we wish to emphasize, however, is that the best scientific appraisal of the cause of an outbreak of illness in an industrial plant should not hinge on whether, for instance, sick leave will be charged for the stoppage.

A word is necessary concerning the role of the psychologist in the investigation of sudden epidemic illness. There is an allegation often made when the psychologist comes in, people refuse to talk anymore. The heart of these statements rests on a belief on the part of some of the public that psychologists only study bizarre phenomena or "nutty" people. Some professionals in the industrial hygiene field reinforce this folk belief with observations and comments that when the psychologist comes in, the case is blown. Both laymen and professionals are laboring under a misconception, the notion that all psychologists are clinicians. In fact, they are not. Just as people understand that all physicians are not psychiatrists, it should be clear that many psychologists, perhaps even most psychologists, are experimentalists interested in how humans perform and behave. They do not deal in psychopathology or observations of any particular person's behavior. They are not there to pronounce MPI victims either ready for the funny farm or ready to go back to work, and they have no interest in either classifying, labeling, or discussing psychopathology.

These psychologists are interested in finding out under what circumstances certain behaviors occur, what conditions exacerbate it, and what conditions ameliorate it. I mention this in particular because there are many of us who have been concerned not so much with only the Mass Psychogenic Illness process itself but also with the affects of MPI-like events such as stress, stressors, tensions (whether physical or social), and their effects on human performance. It may very well be that not only do victims get sick with MPI, but even in circumstances where there is no precipitating factor, no outbreak, no contagion, or no epidemic, people may be experiencing the same problems at subthreshold levels. If people are not performing up to par and that is the result of the same conditions that in some circumstances trigger off MPI, this may suggest productivity losses in a larger variety of circumstances than now envisioned. It would be to everybody's benefit—industry, unions, workers, health workers, hygienists, or whatever, if some amelioration of these conditions could be obtained. Then perhaps, we could be in a Pareto-optimal situation, in which some people are better off, and no one is worse off. Such a result would be more productivity, less tension, fewer sick workers.

Another set of misconceptions that should be put to rest is a set of beliefs concerning what psychological research consists of in an industrial setting. Over the past 4 years I have had the opportunity to hold discussions with a variety of physicians on a number of medical issues. It turns out that they too have, for reasons that I can understand, a surprising notion of what psychologists do. For the most part, they believe that when a psychologist comes in to do research, the study consists of handing everybody an MMPI (Minnesota Multiphasic Personality Inventory), and concluding, for example, that the behavior in question occurs only to people who are high on the hysteria scale and low on the dominance scale.

Firstly, although this type of study (and conclusion) is performed in some cases, it is not typical of most psychological research. As a matter of fact, it is a research strategy that is particularly ill-suited to finding situational changes from one time to another. Instruments such as the MMPI are clinical instruments for diagnosing long-term psychopathology and classifying people for long-standing dispositional or constitutional predilections. These instruments simply do not work very well in recording changes from day to day or week to week or month to month or incident to incident. We have very few good instruments that do that type of task efficiently. I wish we had more, but it is clear that the instruments that are used for psychiatric diagnosis are simply not appropriate and are most often not even used for research in industrial settings. If they are, they are often not very informative. Now, this comment should reemphasize that those psychologists who study MPI are not looking for signs of individual or group psychopathology. This is a free country and if somebody happens to be psychopathological, there is nothing to prevent them from participating, on a free and equal basis, in an MPI outbreak, but it doesn't mean that only people who have psychopathology are enjoined to participate.

SOME UNANSWERED QUESTIONS

There are two major remaining questions concerning the onset of MPI. One is how does an outbreak occur? The embarrassing part about answering that question is it raises the next one: Why doesn't MPI occur more often? There are no really definitive answers to either of these, but let us explore some of those issues.

How does an MPI occur? Well, there are many possibilities. Although we have had a few pioneering studies, the ones by Colligan and his associates (this volume) that review the area and attempt to pull out systematic similarities are really the start of the field. There is so much we don't know that it may be useful as a strategy to advance a simple, easily-explained theory. Such a start is a great help in formulating what research needs to be done and in guiding the future direction of the research. The theory is apt to be wrong. It almost certainly will be in the details, but it undoubtedly will be useful to regard all MPI cases as if they could be explained by a simple sovereign notion. There is no logical reason why all MPI outbreaks have to be triggered off by the same process, just as there is no reason why those illnesses that come from a physical toxin have to come from the same toxin, but for simplicity's sake, because of the consistencies involved, we will assume that the triggering events are fundamentally similar. The trigger event, something that dramatically raises the stress level or tension level or arousal level above what is a threshold value; sometimes it's a crisis event, and sometimes it may be just a chronic buildup of normal levels.

In a previous discussion of MPI, a plant was discussed in which there were two outbreaks of illness, the first of which was attributed to a combination of below-threshold air pollutants that produced a series of physical symptoms. An outbreak a few weeks later with the same symptomotology was attributed to psychogenic causes. This attribution for the second outbreak was regarded as paradoxical and deplored. But this is a concern with the wrong issue. A more basic problem is a tendency of people, even professionals in the field, to assume that symptomotology is a bedrock of hard fact, when in point of issue it's not. Symptom reporting is constructed out of the victim's social beliefs, selective attention to him or herself, expectations, the desire to understand what it is the interviewer wants, and a sensitivity to the ways in which particular questions are formulated. The information we derive from symptomotology following an epidemic of industrial illness must be interpreted in a variety of ways. It cannot be assumed to be the same kind of symptom obtained as when blood pressure or an hematocrit or any of a number of other physical tests are taken; and even with these measures there are problems. When you ask subjects to report things, their reports are filtered through a screen of subjective beliefs and intentions.

Another question that keeps arising is: Why are women reportedly more prone to MPI than men? There are a variety of explanations, and it may be instructive to review some of them.

One theory is that women are more susceptible because they are more likely to be in the sorts of jobs that tend to produce the conditions leading to MPI, whereas men are not. I find this unconvincing for at least two reasons. Firstly, while it may well be true that women get the short end of the stick in the labor force, there are a number of assembly line jobs in which men work in equally drudgerous conditions as women and yet MPI does not occur among groups of men; and secondly, if we look again, not to the industrial settings, but to the MPI cases which have occurred in schools, it's likely that there are similar predisposing conditions in boys' schools or coed schools as in girls' schools, and yet the MPI school outbreaks tend to occur in all-female schools. It's difficult to equate that with the women-stuck-in-rotten-work explanation. Although there are social and cultural factors, such as those discussed by Dr. Kerckhoff and by Dr. Phoon (this volume), that are plausible within their contexts, there still seems to be a need for a gender based explanation whether of socialization, of biology, or interaction to explain the predominately female susceptibility.

CONCLUSION

Finally, a concluding point. Mass psychogenic illness is a sickness. The victims are sick. They throw up, they faint, they have a variety of symptoms. Irrespective of the etiology of these symptoms, it does not alter the fact that these people are as sick as if you had an identifiable physical toxin precipitating their reactions. If we have a person who has an ulcer or a myocardial infarction, that person has a hole in the stomach or some heart muscle dying. It becomes somewhat irrelevant to inquire as to what produced it when it's time to treat it. And to say to that person "It's all in your head" is to add insult to injury. The insult is there and the tendency to utilize a phrase applied to MPI by Dr. Kerckhoff—"to blame the victim" is something that should be resisted, not only because it's unfair to them but because it's probably incorrect.

To reiterate, I believe psychogenic illness does occur. I think it probably occurs more frequently than we expect because we use it mainly as an explanation of last resort. It probably occurs more frequently than we tabulate because it probably also occurs in conjunction with definable physical progenitors of illness; we tend only to count "pure cases" of psychogenic illness as an explanation of sudden epidemic illness. If we looked at "not beyond reasonable doubt" explanations of when MPI occurs, but at plausible occurrences, we would probably find it as a contributing factor in industrial illness in a greater variety of settings than we are now willing to enter into an accounting.

REFERENCE

Johnson, D. M. The "phantom anesthetist" of Mattoon. A field study of mass hysteria. *Journal of Abnormal and Social Psychology*, 1945, *40*, 175–186.

III THEORETICAL APPROACHES TO MASS PSYCHOGENIC ILLNESS

Investigating a phenomenon like mass psychogenic illness is a little like sky diving in that if you don't get it right the first time, you rarely get a second chance. The occurrence of such illnesses are relatively rare and unpredictable, creating considerable confusion and anxiety while they last, and guarded relief when they are over. The investigator has precious little time to decide what data to collect, and how best to collect it. Obviously some theoretical structure and guidance is extremely helpful in focusing attention on relevant parameters of the situation and steering us away from probable dead ends. Equally important is the role that theory can play in tying mass psychogenic illness more closely to conventional psychosocial thought. Models of attribution and social comparison, conformity and contagion, social learning, coping, rumor transmission, communication, attitude formation and change, interpersonal perception, and similar traditional psychological processes have a direct bearing on our understanding, containment, and prevention of contagious psychogenic illness. In fact, a general conclusion of this volume seems to be that mass psychogenic illness is a *social* phenomenon affecting a proportion of *normal* population under certain physical and psychosocial conditions.

The papers in the following section discuss a variety of theoretical approaches to components of the illness episode. In addition to their immediate contribution to our understanding of mass psychogenic illness, they provide interesting case studies of the application of social psychological, perceptual, sociological, and even psychodynamic theory to practical problems. The seven chapters within this section approach the mass psychogenic illness phenomenon from different perspectives and levels of analysis. The first, by Dr. Pennebaker, deals with some of the perceptual processes implicated in the awareness of physical symptoms during illness outbreaks. Drs. Singer, Baum, Baum, and Thew discuss mass psychogenic illness within the context of social comparison theory. Dr. Freedman examines theories of contagion from several psychological perspectives. In Chapter 11, Dr. Stahl introduces the concept of emergent norm as a way of explaining some of the social processes surrounding mass psychogenic illness. Dr. Kerchkoff's chapter is concerned with several social psychological and sociological factors that are implicated in the contagion process. In Chapter 13, Dr. Sirois offers a psychodynamic account of the phenomenon wherein he draws upon a large number of cases that have occurred over the centuries. Finally, Drs. Folkman and Lazarus discuss their coping theory as an explanation of the mass psychogenic illness phenomenon.

The various theoretical approaches to mass psychogenic illness should be viewed as complementary as opposed to competing. Each deals with a different facet of the phenomenon. This cross-fertilization of laboratory and field research can only help to enlighten our understanding of mass psychogenic illness.

8 Social and Perceptual Factors Affecting Symptom Reporting and Mass Psychogenic Illness

James W. Pennebaker
University of Virginia

ABSTRACT

Evidence is presented indicating that the perception of physical sensations and internal state parallels the processes involved in the perception of external environmental events. Two factors influencing the awareness of internal state are presented. If the external environment is lacking in information, the probability of attending to internal sources of information increases. In addition, the schemas or hypotheses we adopt about our health can result in highly specific monitoring of internal sensations. Further, individuals are much more likely to become aware of sensations that can verify an illness schema than sensations that could disconfirm such a schema. Ways in which illness schemas are established and result in selective search of internal state are presented. Finally, implications for the understanding and prevention of mass psychogenic illness episodes are discussed.

INTRODUCTION

A fundamental issue in understanding psychogenic disorders concerns how individuals perceive internal sensations and physical symptoms. The present work addresses when individuals are most likely to attend to sensations and how these sensations can be interpreted. This approach is based on a perceptual perspective. That is, the processes that have traditionally dealt with the perception of external environmental events can also be applied to the perception of internal state.

In many respects, the perceptual approach supplements many of the perspectives that are advanced in this volume. Social comparison, coping, and other theories often begin with the assumption that individuals are aware of internal arousal or physical sensations. I am concerned with the factors influencing the initial awareness of the sensations themselves. In other words, the perceptual approach is limited to persons' experiences of their sensations on a more molecular level. The subsequent contagion of the sensations or symptoms is more directly addressed by many of the broader social influence theories.

Much of the work presented is the product of several experiments my students and I have conducted over the last few years. Many of the studies have dealt with relatively minor symptoms—coughing, itching, nasal congestion, perceptions of cold hands and racing hearts. Obviously, ethical considerations have limited the domain of symptoms that we can study in the laboratory. However, our findings concerning the various mundane complaints should generalize to more serious physical problems. One reason we are confident about the generalizations of our findings is that we have found divergent symptoms to be influenced in similar ways by common perceptual factors. For example, perceptions of nasal congestion, heartrate, itching, etc. can be manipulated and influenced in comparable ways using similar instructions. In other words, the perception of divergent physiological symptoms—whether under the control of sympathetic, parasympathetic, or even the central nervous systems—is highly dependent on attentional and interpretive factors. Consequently, a paradigm that can promote itching or nasal congestion could, in theory, also increase vomiting, fainting, or other more serious physical problems.

The present chapter is divided into three sections. The first discusses the perceptual process in general. The second section deals with attentional and interpretive factors implicated in the perception of internal sensations. The final section will link the perceptual approach to the contagion of psychogenic disorders and recommend techniques that should be effective in limiting their occurrence.

PERCEPTION OF EXTERNAL AND INTERNAL EVENTS

All organisms seek to learn about their worlds. In this learning process, we take in information about objects, events, etc. Although this information is apprehended by visual, auditory, or other receptors, it must be processed at a higher level within the brain. Perception represents a highly active and cyclical process. We do not passively wait for information or stimuli to bombard our receptors and brain. Rather, we actively and selectively seek information.

Many theorists and researchers have addressed how the perceptual process works in dealing with the external environment (e.g., Gibson, 1966; Lazarus, 1966; Neisser, 1976; Newtson, 1976). We can study the process from two perspectives. The first is by looking at how external stimulus information is attended to and encoded into categories, schemas, sets, or tentative hypotheses. This view is unidirectional from external events to our understanding of them. The second approach concerns how these categories, schemas, etc. direct attention or search behavior for external information. Clearly, we do not encode everything in our visual world. Instead, we selectively look for information based on our expectations. Our selective search behavior, then, restricts the type of information that we may glean from the environment.

An example of this process can be seen in our walking in the woods and hearing a rustle in some bushes. The rustle may draw our attention because of its novelty, motion, relative complexity (e.g., Berlyne, 1960), or because there are no other competing stimuli present at the time. We may then develop a tentative schema or hypothesis such as "Is it a snake?" Given such a schema, we may selectively look at the base of the bushes for a long, slithering object. Another equally plausible hypothesis, i.e., "Is it the wind?" will result in our looking for different information (leaves shaking, sensation of coolness on skin, etc.). The type of information that we find may confirm or disconfirm our schema or hypothesis. If it confirms the snake hypothesis, we may then form new sets and begin looking for new information (e.g., is it poisonous, can I run away or cope effectively?). In sum, the perception of external events is an ongoing process involving both taking in external information *and* creation or modification of schemas which direct our search behavior. To understand the process, we can study factors that affect the initial awareness of the stimulus as well as the schemas we adopt and the subsequent search behavior deriving from the schemas in order to learn more about the stimulus.

Of particular relevance to the present research are factors that can bias or distort the information we gain from stimuli. One problem is the role of competing stimuli in our noticing a given stimulus. For example, we are more likely to notice a rustle in the bushes if we are sitting alone in the quiet woods than if we are actively engaged in a conversation with a fellow camper. In other words, the probability of noticing any given stimulus is inversely related to the number of competing stimuli in the environment at the time. In addition, recent evidence by perceptually-oriented social psychologists indicate that the schemas that we hold will restrict the information that we can glean from the environment. Specifically, we are much more likely to find schema-consistent or relevant information than schema-inconsistent information (cf., Crocker & Taylor, 1978; Snyder, 1979). Using the above example, if we adopt the snake hypothesis, we would be more likely to notice a long object at the base of the bushes and believe it to be a snake (even if it were a stick), than if we did not initially hold the snake hypothesis. In addition, the snake hypothesis would restrict our encoding information that the wind had caused the rustle.

A basic premise of this chapter is that the perception of internal sensations represents the same processes of perception implicated in our learning of external events. Individuals are more likely to become aware of physical sensations at some times than others. Further, the schemas or sets to which we have access will guide our search for particular sensations. Such a perceptual view has several implications for symptom-reporting in general and incidences of mass psychogenic illness in particular. The following section will be devoted to some recent findings supporting this perceptual approach. Evidence is presented demonstrating how lack of external information will increase the probability of becoming aware of internal sensations. Secondly, data supporting the contention that illness-relevant schemas result in the selective encoding of internal sensations is offered.

ATTENTION, SCHEMAS AND SELECTIVE SEARCH: RESEARCH EVIDENCE

In order to experience and report physical symptoms, individuals must attend to their bodies. As discussed above, becoming aware of internal stimuli will be dependent, in part, on the nature of competing external information. In addition, the schemas we hold concerning our bodies will greatly influence our awareness of sensations via selective search. For future reference, schema, as used here, is defined as a hypothetical entity that readies the perceiver for perceptual experience by providing a structure for assimilating new information. Hence, an illness schema will prompt the perceiver to selectively search his/her body for illness-relevant sensations. The present section is divided into two parts. The first concentrates on factors that will increase the probability of attending to internal sensations. Before we can define sensations as illness related, we must first become aware of them. The second part focuses on the role of selective search as a function of the schemas we hold. This section will explore both selective search and factors that can establish an illness schema in the first place.

Competition of External and Internal Cues

Historically, perceptual psychologists have been interested in why we are more likely to attend to certain stimuli than others. For example, Berlyne (1960) notes that stimuli that are novel, complex, or display motion will be examined in greater detail than redundant, simple or stationary arrays. If we assume that attention or search can alternate between external and internal stimuli (cf., Wicklund, 1975), the probability of attending to internal sensations will depend, in part, on the lack of external information and/or heightened salience of the physical cues themselves. In other words, given that both external and internal sources of information are potentially available at most times, the individual must selectively process data. Given the high number of available sources of information, the processing of any given internal or external stimulus will restrict the encoding of competing stimuli. This implies that external information can be used at the expense of internal information and vice versa.

Logically, in order to experience a sensation, the individual must attend to the body. Several demographic studies indicate that if persons' living or working environments are not stimulating, they are much more likely to report physical symptoms. For example, individuals living alone or in a rural area or in an environment characterized by social isolation report a heightened number of symptoms and claim to be in worse health (Moos & Van Dort, 1976; Pennebaker & Funkhouser, 1979; Wan, 1976). Work environments perceived as repetitious and boring are positively related to health center and sick call visits (Greenley & Mechanic, 1976; Moos, 1975), and often typify incidences of mass psychogenic illness (Colligan & Murphy, 1979; Stahl & Lebedun, 1974). Comparable findings have been reported by person-environment fit researchers. That is, individuals whose abilities exceed the demands of their jobs evince an increased degree of somatic complaints (Harrison, 1976), symptom reports (Coburn, 1975), and signs of somatic depression (French, 1973). Wright, Kane, Olsen, and Smith (1977) found that smelter and mine workers who found their jobs uninteresting over-reported respiratory symptoms relative to an objective measure of respiratory dysfunction. In other words, lack of external stimulation is associated with reporting of physical complaints. Of course, each of these studies was correlational, thereby making it difficult to establish causation. However, the converging evidence suggests that when external demands are low, the probability of attending to internal sensations increases. Once aware of the sensations, individuals may become more attentive to internal state in general, exaggerate them, and ultimately seek medical attention.

Recently, we have begun to examine health-relevant behaviors as a function of external stimulus demands in experimental settings. The first line of research on this problem has dealt with coughing behavior (Pennebaker, 1980). Coughing is a response to the perception of a tickling or irritated throat. In order to be aware of tickling or irritation, we must attend to our throat. In the first of several studies, we showed an 18 min movie to 50 college students. They were asked to rate the movie on how interesting it was every 30 sec, along a 7-point scale. We then showed the movie to three Introductory Psychology classes of 200, 90, and 40 students. While watching the movie, the number of coughs were counted and collapsed into the same 30-sec blocks previously rated on their interest. The simple correlation between the number of coughs per 30-sec interval and interest rating for the same intervals was a highly significant $-.57$. In other words, the less interesting parts of the movie were characterized by a large number of coughs. Similarly, we have found negative correlations between students' liking of a given course and the number of coughs emited during a typical lecture by the teacher of the course.

A second line of research substantiates this reasoning within a person-environment fit paradigm (Pennebaker & Brittingham, 1981). In the experiments, college students worked on serially-presented simple arithmetic problems at one of three paces: slow, moderate, or fast. As expected, people reported the most physical symptoms (e.g., headache, racing heart, upset stomach) in the slow and fast conditions. In a second study, we sought to tease out factors influencing attention to body and actual autonomic arousal. Again, people worked on the arithmetic problems at one of the three paces. Skin conductance was continuously measured. In addition, subjects were told that they would receive short bursts of air on their arm and back over the course of the study. In fact, they received six squirts on the arm and none on their back. Symptom-reporting was again curvilinear with Moderate-paced subjects reporting the fewest symptoms. Skin conductance and accuracy of recall about the squirts served as measures of autonomic activity and attention to body, respectively. If differential autonomic activity accounts for the curvilinear pattern for symptom reporting, skin conductance should be higher in the Slow and Fast conditions and lowest in the Moderate cell. Conversely, if attention to body underlies symptom-reporting, we would expect greatest accuracy of recall about the squirts in the Slow and Fast conditions. In fact, both processes appear to be working. In terms of recall about the squirts, only subjects in the Slow condition were accurate. The Moderate and Fast subjects were equally poor at recalling the number and location of the squirts. In terms of skin conductance, only Fast subjects maintained a high level throughout the experiment. Both Slow and Moderate subjects adapted equivalently.

These data, then, offer indirect evidence that an environment lacking in external stimulation allows for greater attention to body, thus resulting in increased probabilities for reporting physical symptoms. When a person must deal with an overabundance of external information, he or she evinces heightened arousal or autonomic activity. Once the task is over, the person probably becomes aware of sweating hands, racing heart, etc. and accurately reports them. These data, together with other studies (e.g., Pennebaker & Lightner, 1980) indicate that situations lacking in external information result in attentional focus to the body and subsequent symptom awareness and reporting. The implications of this relationship will be discussed in the final section.

Schemas and Selective Search

The approach just described represents an unguided search for information. That is, at certain times individuals have an increased probability of noticing internal sensations. Although our search for information about our bodies is often dependent on lack of external information, much of the time we are directly attempting to learn about specific sensations based on hypotheses, labels, or schemas we hold. These schemas are particularly important in that they result in highly specific attention to certain sensations. Further, the sensations will be experienced in very different ways as a function of the schemas.

Before discussing our research concerning selective search, it may be instructive to report a case study related to contagion that occurred in a summer school class I taught in 1978. The class of 25 students met from 10:30 to 1:30 each day in a biology laboratory classroom. The large room was equipped with gas outlets. On one day the air conditioner was broken, causing the room temperature to increase to about 85° F. Both the students and I became extremely tired as my lecture droned on. Had I passed out a symptom checklist, virtually everyone would have reported that they felt drowzy and sweaty because of the heat. In other words, our perception of the heat resulted in our monitoring specific heat-relevant sensations. Midway into the lecture, a woman on the first row raised her hand and announced that she smelled gas escaping from one of the jets. Immediately, I noticed that I felt nauseous, dizzy, and on the verge of fainting. These perceptions were shared by over half of the students. (In fact, those who did not experience these symptoms assumed that I had set up the entire scenario as a class experiment!). As an aside, there was no escaping gas.

This minicontagion episode exemplifies the role of selective search as a function of available schemas. First, the heat had produced a variety of common but ambiguous sensations. Having experienced heat, we all were familiar with certain sensations that accompany it. In a hot room, then, we selectively attend to parts of our body that are relevant—sweating, drowziness, etc. We typically do not look for irrelevant sensations. However, the introduction of a new schema or label will cause us to monitor other sensations that may be consistent with the new schema. In other words, we are selectively gathering new sensory information and organizing it differently. From the example, switching from a heat to an asphyxiation mode causes us to interpret drowziness as faintness, or hunger pains as nausea. We ignore our heat-induced sweating and other asphyxiation-irrelevant sensations. In sum, the adoption of a new schema will result in our attending to different sensations and interpreting them differently.

This assumption has prompted two related lines of research. The first concerns *how* schemas result in the selective search and encoding of sensations. The second question deals with what factors can establish differing types of body-relevant schemas. These two problems will be discussed separately.

Selective Search

For the past year, J. A. Skelton and I (Pennebaker & Skelton, 1978; 1981) have conducted several experiments on factors influencing the perceptions of nasal congestion, heartrate, and skin temperature. Our initial idea was that merely focusing attention on some part of the body would result in heightened symptom reports. In the first experiment dealing with nasal congestion, subjects were asked to pay attention either to their nose and nasal passages or to an irrelevant object. To our surprise, we found no differences in perceived nasal congestion following these attention instructions. Our next tack was to manipulate *how* people were to attend to nasal sensations. That is, we sought to establish congestion-relevant schemas so that the nasal sensations could be organized in a more specific way.

The second experiment manipulated focus of attention and the ways in which nasal sensations could be encoded. Subjects were randomly assigned to one of four experimental conditions. As in the first study, two of the conditions required subjects either to attend to their nose and nasal passages (Nose Neutral condition) or to an irrelevant object (Distraction condition). In addition, one group of subjects was instructed to attend to sensations indicating "congestion, nasal blockage, and difficulty in breathing through the nose" (Nose Negative). Finally, a fourth group of subjects attended to sensations indicating nasal clarity and air passing freely through the nose (Nose Positive). The Nose Negative and Nose Positive subjects, then, had instructions to selectively attend to nasal sensations. As can be seen in Table 1, perceptions of congestion were influenced only in the two selective attention cells. Following the experiment, each subject was extensively debriefed. Three subjects—all in the Nose Negative condition—blew their nose during this time. No others did so. One subject in the Nose Positive cell remarked that her nose hadn't felt so clear in weeks despite her using various decongestants. Unfortunately, the nasal congestion study did not employ an actual measure of congestion. To learn if the schema related instructions actually produced somatic change, a third experiment was conducted.

The next study examined changes in perceptions of heartrate and employed an actual measure of heart beat. As in the Nose study, subjects were instructed to attend either to their heartrate (Neutral condition), a racing or accelerating heartrate (Acceleration condition), or a slowing or decelerating heartrate (Deceleration). Although there were no differential changes in actual heartrate, Acceleration subjects perceived an increase in their heart rate, Deceleration subjects perceived its slowing down, and Neutral subjects perceived no change (means of the changes of the three groups: Acceleration = +11.0; Deceleration = −9.4; Neutral = −1.9; all groups were significantly different from each other at $p < .05$, except for the Deceleration versus Neutral contrast which differed at p

TABLE 8.1
Self-Reports of Nasal Congestion Before and After the Attention Manipulation

Condition	*Before*	*After*	*Difference*
Nose Neutral	60	67	$+7^{ab}$
Nose Negative	56	77	$+21^{a}$
Nose Positive	74	61	-13^{b}
Distraction	45	45	0^{ab}

Note: Responses were along a 100 mm line, where 1 = not at all congested, and 100 = completely congested. N = 12 per cell. Numbers with different superscripts differ at $p < .05$.

= .15). The results of this study, then, indicate that how subjects attend to their internal sensations is crucial in bringing about changes of perceptions of physical symptoms. Further, significant changes in perception are not entirely dependent on physiological change.

A clear problem with the above studies is that we were placing heavy demands on the subjects to report sensations in one way or another. In addition, subjects could have distorted their self-reports knowing that we could not detect if they were accurate or not. Even in the Heartrate study, subjects were only told that we were measuring some physiological index and were not told that it was a measure of heartrate. Consequently, we conducted a fourth experiment looking at changes in skin temperature using a different paradigm. In the study, subjects were told that we were examining the aftereffects of ultrasonic noise on task behavior. Subjects expected to listen to a high frequency sound beyond their hearing range while we monitored skin temperature. In addition, we told the subjects that the (bogus) noise had been shown to cause vasodilation or warming of the hands, vasoconstriction or cooling, or have no effect on skin temperature. Although all subjects thought they were listening to ultrasonic noise over headphones, they actually heard nothing. Following the "noise" period, they completed a short questionnaire ascertaining the degree to which they noticed their finger's becoming warmer or cooler. As expected, Increase subjects detected their finger's becoming warmer, Decrease subjects reported a drop in temperature, and Control subjects detected no change. (Means of the three groups along a 100 mm line, where 1 = becoming cooler, and 100 = becoming warmer: Increase = 69.6; Decrease = 48.1; Control = 57.8. All means were significantly different from each other at $p < .05$.)

As in the Heartrate study, there were no differential changes in actual skin temperature. Although self-reports of change were not correlated with actual change, an interesting pattern of results emerged in examining fluctuations in skin temperature. A fluctuation was defined as a change in the direction of temperature during the noise period, such as temperature becoming warmer, then cooler, or vice versa. Simple within-cell correlations of fluctuations with self-reported skin temperature were +.48 in the Increase cell, −.49 in the Decrease cell, and a non-significant +.11 in the control condition. These data indicate that subjects were only encoding or noticing changes that would verify their expectations and ignoring disconfirming skin temperature information.

Taken together, these studies demonstrate that merely focusing attention on the body is not a sufficient condition for changes in the perception of physical symptoms or sensations. Rather, people must have a tentative schema, hypothesis, or general way of organizing and encoding internal sensations for them to have meaning. In other words, the schema guides ones' focus of attention, resulting in the selective encoding of sensations. Further, this search is biased in that one is more likely to encode schema-verifying information than information that could disconfirm the schema.

Given that the schemas we hold are important factors in the perception of internal state, it is critical to examine how schemas are adopted. That is, under what conditions are we most likely to encode internal sensations within an "illness" or perhaps "mass psychogenic illness" mode? Althought there are a variety of developmental, cultural and dispositional variables implicated in this process, I restrict my discussion to social factors that can establish body-relevant schemas. It must be noted, that at this point many of the approaches that are presented in this volume directly bear on the factors that can induce an illness schema. Although our terminologies may differ, we are addressing the same factors.

Establishment of Illness Schemas. For the past several months, my students and I have been attempting to discern which factors are most likely to result in a person's selectively attending to the body. Examination of the mass psychogenic illness literature clearly indicates that seeing another person become sick or emit a bizarre behavior will increase the probability that others will behave likewise (cf., Markush, 1973). In other words, unambiguous overt behavior is more likely to spread than highly subtle illness behavior. An example of this is coughing behavior. In analyzing coughs during exams, we have found that coughs occur in "bunches." That is, there may be 30 sec of silence and then 8–10 coughs in the space of 5 sec. Just hearing another person cough, then, results in a quick monitoring of one's throat and the emission of a cough (Pennebaker, 1980).

One of our first questions concerning the establishment of a health or illness schema dealt with the probable cause of the illness. Recall that in many of the recent contagion incidences there has been a presumed cause—an invisible June Bug (Kerckhoff & Back, 1968), a mystery gas (Stahl & Lebedun, 1974), pollutants in the air (Rouche, 1978), etc. Is the attribution of a cause for the illness a necessary condition for the immediate spread of the symptoms? That is, after seeing a coworker collapse, does the individual need to go through the cognitive process: "Joe just collapsed; he must have come into contact with some toxic agent that caused it; I have probably also been exposed to the toxic agent; therefore I, too, will collapse." Some preliminary work suggests that this attribution process does not always occur.

Our first study examined scratching behavior, which we can assume is a response to an itching sensation. We were interested in learning the degree to which a possible cause of an itch would result in another person's also scratching. In the study that took place in various libraries on campus, two experimental confederates sat down by a person who was reading. After one min, the confederate sitting next to the unsuspecting subject did and said one of four things: she began scratching and announced to the other confederate (in a stage whisper) that she was just bitten by a mosquito (Mosquito condition); she scratched and said that she got sun poisoning at the lake (Sun condition); she scratched and said she was hungry and wanted to go get a hamburger (No Cause condition); she did *not* scratch and said she wanted to get a hamburger (No Scratch). The two confederates then left while an observer watched the subject for one minute to see if she or he scratched. In sum, the confederate scratched and gave a possible contagious cause (mosquitoes), an implausible contagious cause (sun poisoning), or no cause at all. Of the 16 subjects in each condition, 50% scratched in the Mosquito and Sun cells, 32% in the No Cause cell, and nobody scratched in the No Scratch condition. In other words, people's scratching was unrelated to the purported cause of the confederate's itching.

In retrospect, these results are not surprising. For example, when my research assistants and I got together to talk about the study, we would all catch ourselves scratching feverishly even though there was no objective agent causing the itch. In various interviews, people note that if they see someone vomit, they too become nauseous. In other words, merely observing illness behavior sets up a tentative schema or hypothesis such as "Am I sick? or Do I itch somewhere?" This schema results in specific monitoring of relevant sensations.

It must be noted that such schemas can be established in ways other than observing overt behavior. For example, reading or hearing about an illness can prompt search behavior. Audrey Burnam and I (Burnam & Pennebaker, 1977) conducted two simple questionnaire studies examining the verbal encoding of schemas. In one study, we simply passed out a symptom questionnaire that was headed in one of two ways. One version—the Illness Set—simply noted, "As you know this is the time of the year when we are surrounded by cold- and flu-producing viruses as well as other illness-related organisms." The other version omitted the statement. As we predicted, people reported significantly more physical symptoms in the Illness Set condition than the No Illness Cell. This trend was especially strong for the flu-relevant symptoms (e.g., headache, upset stomach) than the symptoms unrelated to the flu (e.g., itching, ringing in ears). These data, then, indicate that illness schemas can also be established along modes other than overt behavior.

A particularly striking example of the establishment of illness schemas via verbal encoding can be seen in cases of medical students' disease wherein medical students adopt the physical symptoms of the diseases that they are studying (cf., Hunter, Lohrenz, & Schwartzman, 1964). Interestingly, as in cases of mass psychogenic illness, most incidences of medical students' disease occur when individuals are under great stress. That is, ambiguous internal sensations associated with heightened autonomic activity can readily be encoded into the available illness schemas.

Summary of Empirical Findings

This section broadly addressed what factors can influence the awareness and reporting of physical sensations. Individuals constantly seek information about their world. If the external environment is lacking in meaningful information, individuals are more likely to attend to internal, sensory sources of information.

Attention to internal sensations is also influenced by the schemas we adopt. These schemas can result in monitoring of selective sensations to confirm or disconfirm the schemas. The search process is biased in that people are more likely to notice schema-relevant sensations than sensations that could disconfirm the schema. Finally, the schemas we adopt about our bodies can derive from seeing, hearing, or reading about illness. Further, the adoption of an illness schema need not be processed in a causal way. That is, individuals do not necessarily go through the process of assuming that the cause of a coworker's illness is the same cause of their own sensations. The following section will address some implications of the perceptual perspective for the contagion of psychogenic disorders.

IMPLICATIONS FOR MASS PSYCHOGENIC DISORDERS

As has been stressed, symptom awareness requires attention to body. Any factors that promote monitoring of internal state, then, should increase the probability of perceived health problems which may result in higher absenteeism rates or cases of psychogenic contagion. A job that is undemanding because of its repetitiveness, lack of intrinsic interest, or even slow pace may discourage external attentional focus. Lack of social interaction should also promote this process (cf., Cobb, 1976; Colligan & Murphy, 1979). Social support has the dual function of drawing attention away from the body as well as providing an information source for the person to better evaluate the sensations. Persons in settings lacking external information or social support have more time to notice and dwell upon subtle sensations. The more they inspect these sensations, they may begin to worry and define them in illness terms.

In practice, the establishment of environments that encourage external focus of attention may be difficult to engineer. Depending on demands of a job, music or other distractors may be feasible. However, such a tack may only be effective in the short term. Ideally, of course, boring jobs should be made more interesting, some degree of social interaction should be encouraged, pacing of tasks should take the persons' abilities into consideration. Hopefully, management and labor could be told about the attentional factors involved and work out solutions that would be practical. If adjustments in the work setting could be accomplished, the payoff for all involved would be high.

The problems of illness schemas and selective search pose a more basic concern. Our research indicates that merely observing or hearing about an illness is likely to lead to an immediate search for comparable sensations. Although hypothesized physical agents such as odors, bug bites, etc. may prolong the contagion episodes, they may, on occasion, be secondary factors in the initial stages. That is, seeing someone vomit or faint may lead to an immediate monitoring of internal sensations. The assignment of a causal agent may well occur *after* the person has also detected signs of nausea rather than before.

Virtually every episode of mass psychogenic illness occurs among workers or students who are tense, anxious, overworked, or in a state of increased autonomic activity. As has been suggested, this arousal could be the result of poor worker-management relations (Colligan, 1978), poor lighting or temperatures (Colligan & Urtes, 1978), crowding (Freedman, 1979), loud noise (Shepard & Kroes, 1975), role or status incongruities (Kerckhoff & Back, 1968), family or financial pressures (Colligan & Murphy, 1979), or a combination of these and other stressors. No matter what the true cause of the arousal, the important fact is that the affected individuals are constantly experiencing a high number of diverse and ambiguous internal sensations. Based on our research with illness schemas, it would be predicted that the more physical sensations present, the easier it would be to confirm any number of illness hypotheses. That is, with such a high number of available sensations to monitor, a person could easily confirm his or her being asphyxiated, succumbing to the flu or more exotic ailment such as radiation poisoning. As an aside, the recent nuclear reactor problems may serve as particularly fertile bases for future illness schemas. As Robert Lifton (1967) has chronicled in his book concerning the Hiroshima victims, and as recent news media accounts surrounding the Three Mile Island accident imply (cattle deaths, illnesses, etc.), an ambiguous illness schema such as radiation can essentially account for *any* physical symptoms. Consequently, individuals exposed to radiation may well chronically attend to internal sensations at a much higher rate than if they were not exposed. Although this may not directly result in cases of mass psychogenic illness, it could have the effect of greatly increasing absenteeism and visits to medical personnel.

Obviously, the best way to circumvent the selective search process would be to reduce the true causes of anxiety and arousal. This would effectively diminish the number of sensations a person could use to confirm a given illness schema. Given the real world, however, the reduction of tension is not always a viable alternative. In such cases, an effective preventive technique may be to establish or alter peoples' schemas concerning their sensations. A possible way of accomplishing this would be to teach individuals which physical sensations are typically associated with tension, noise, or job-related stress. From my own talking with hundreds of undergraduate students, it is clear that most do not always associate stress with specific constellations of symptoms. We have found that most people usually associate only one or two sensations with tension—even though our physiological measures indicate they are evincing changes in skin conductance, heartrate, or skin temperature. In other words, when people are in an anxiety-provoking situation, they are attending to only a select group of possible sensations. In so doing, they are ignoring other tension-relevant sensations which could potentially verify other competing schemas. If individuals were trained to associate specific groups of sensations to the autonomic changes associated with their jobs, they might be less apt to later attribute them to illness.

A final issue is how to deal with cases of mass psychogenic illness once they have begun. Simply telling the affected individuals that their symptoms are not really occurring or are purely psychosomatic would not be effective. They *are* experiencing the symptoms. They do have concrete sensory data to verify their illness schemas. Unfortunately, the attending physicians, psychologists, NIOSH researchers, or other personnel are in a major bind since, in some cases, the contagion may truly be the result of toxic agents rather than misperceptions of internal state. Consequently, only after medical or environmental tests have indicated no physiological bases for the disease can the workers be appraised of the possible psychological causes. Once aware of the perceptual processes implicated in the contagion episodes, similar outbreaks within the job or school setting may be prevented.

REFERENCES

Berlyne, D. *Conflict, arousal, and curiousity.* New York: McGraw-Hill, 1960.

Burnam, M., & Pennebaker, J. *Cognitive labeling of physical symptoms.* Eastern Psychological Association, Boston, 1977.

Cobb, S. Social support as a moderator of life stress. *Psychosomatic Medicine,* 1976, *38,* 300-314.

Coburn, D. Job-worker incongruence: Consequences for health. *Journal of Health and Social Behavior,* 1975, *16,* 198-212.

Colligan, M. *An investigation of apparent mass psychogenic illness in a furniture plant.* Unpublished NIOSH evaluation report, 1978.

Colligan, M., & Murphy, L. Mass psychogenic illness in organizations: An overview. *Journal of Occupational Psychology,* 1979, *52,* 77-90.

Colligan, M., & Urtes, M. *An investigation of apparent mass psychogenic illness in an electronics plant.* Unpublished NIOSH evaluation report, 1978.

Crocker, J., & Taylor, S. Theory driven processing and the use of complex evidence. *American Psychological Association.* Toronto, 1978.

Freedman, J. *Theories of contagion.* Paper presented at American Industrial Hygiene Conference, Chicago, 1979.

French, J. Person role fit. *Occupational mental health,* 1973, *3,* 15-20.

Gibson, J. J. *The senses considered as perceptual systems.* Boston: Houghton Mifflin, 1966.

Greenley, J., & Mechanic, D. Social selection in seeking help for psychological problems. *Journal of Health and Social Behavior,* 1976, *17,* 249-262.

Harrison, R. *Job stress as person-environment fit.* American Psychological Association, Washington, 1976.

Hunter, R., Lohrenz, J., & Schwartzman, A. Nosophobia and hypochondriases in medical students. *Journal of Nervous and Medical Disease,* 1964, *139,* 147-152.

Kerckhoff, A., & Back, K. *The June bug: A study of hysterical contagion.* New York: Appleton, 1968.

Lazarus, R. *Psychological stress and the coping process.* New York: McGraw-Hill, 1966.

Lifton, R. J. *Death in life.* New York: Randon House, 1967.

Markush, R. E. Mental epidemics: A review of the old to prepare for the new. *Public Health Reviews,* 1973, 353-442.

Moos, R. *Evaluating correctional and community settings*. New York: Wiley, 1975.

Moos, R., & Van Dort, B. *Student physical symptoms and the social climate of college living groups*. Unpublished manuscript, Social Ecology Laboratory, Department of Psychiatry and Behavioral Sciences, Stanford University, 1976.

Neisser, U. *Cognition and reality*. San Francisco, 1976.

Newtson, D. Foundations of attribution: The perception of ongoing behavior. In J. Harvey, W. Ickes, & R. Kidd (Eds.), *New directions in attribution research, Volume 1*. Hillsdale, N.J.: Lawrence Erlbaum Assoc., 1976.

Pennebaker, J. W. Perceptual and environmental determinants of coughing. *Basic and Applied Social Psychology,* 1980, *1,* 83–91.

Pennebaker, J. W., & Brittingham, G. L. Environmental and sensory cues affecting the perception of physical symptoms. In A. Baum & J. Singer (Eds.), *Advances in Environmental Psychology, Volume 4*. Hillsdale, N.J.: Lawrence Erlbaum Assoc., 1981.

Pennebaker, J. W., & Funkhouser, J. E. *The influences of social support, activity, and life change on use of medication among an elderly population*. Unpublished manuscript (mimeo). University of Virginia, 1979.

Pennebaker, J. W., & Lightner, J. Attentional focus and perceptions of internal state. *Journal of Personality and Social Psychology,* 1980, *39,* 165–174.

Pennebaker, J., & Skelton, J. Psychological parameters of physical symptoms. *Personality and Social Psychology Bulletin,* 1978, *4,* 524–530.

Pennebaker, J. W., & Skelton, J. A. Selective monitoring of physical symptoms. *Journal of Personality and Social Psychology,* 1981, *41,* 213–223.

Roueche, B. Untitled article. *New Yorker,* August 21, 1978, 63–70.

Shepard, R. D., & Kroes, W. H. *Report of an investigation at the James Plant*. Internal report prepared for the National Institute ofr Occupational Safety and Health. Cincinnati, Ohio, 1975.

Snyder, M. Self-monitoring processes. In L. Berkowitz (Ed.), *Advances in experimental social psychology* (Volume 12). New York: Academic Press, 1979.

Stahl, S., & Lebedun, M. Mystery gas: An analysis of mass hysteria. *Journal of Health and Social Behavior,* 1974, *15,* 44–50.

Wan, T. Predicting self-assessed health status: A multivariate approach. *Health Services Research,* 1976, *11,* 464–477.

Wicklund, R. A. Objective self-awareness. In L. Berkowitz (Ed.), *Advances in experimental social psychology,* vol. 8. New York: Academic Press, 1975.

Wright, D., Kane, R., Olsen, D., & Smith, T. The effects of selected psychosocial factors on the self-reporting of pulmonary symptoms. *Journal of Chronic Disease,* 1977, *30,* 195–206.

9 Mass Psychogenic Illness: The Case for Social Comparison

Jerome E. Singer
Uniformed Services University of the Health Sciences

Carlene S. Baum
Human Design Group—Olney, Maryland

Andrew Baum
Brenda D. Thew
Uniformed Services University of the Health Sciences

ABSTRACT

There are consistencies among the various outbreaks of Mass Psychogenic Illness (MPI) that show resemblances to the general symptoms of stress—a group of diffuse, undifferentiated changes in arousal and internal state. Social comparison, a theory developed to explain how people identify and evaluate their opinions, abilities and emotions, is described and suggested as a mechanism for the instigation and propagation of MPI. The specific features of MPI outbreaks, and the characteristics of the affected workers are examined in a social comparison framework. The question of why women are so overrepresented among those affected is also discussed.

The occurrence of mass psychogenic illness (MPI) has been traditionally explained in a variety of ways, ranging from contagion and group influence analyses to consideration of social isolation (e.g., Kerckhoff et al., 1965). This chapter considers outbreaks of MPI from a different perspective, using research and theory on social comparison processes, affiliation, and evaluation of emotional states to derive hypotheses about the etiology of MPI. A general framework for this view will be developed, an analysis of the fit between this framework and the existing data will be made, and implications for experimental explorations of MPI will be discussed.

The literature on MPI is sometimes confusing and unsystematic. This occurs, in part, because of the nature of MPI; it is not only a spontaneous and relatively short-lived phenomenon (making prospective studies costly and difficult) but it is also seen at the time of its occurrence as caused by an external pathogen. Thus, it is only *after* medical or hygiene personnel have determined that the outbreak of illness has not been caused by a toxin that psychogenic considerations are explored (e.g., Colligan & Murphy, 1979).

Although the occurrence and symptomatology of MPI have varied, certain consistencies across most outbreaks have been identified. In their review of MPI in organizations, Colligan and Murphy (1979) have listed five factors that seem to be characteristic of work environments in which outbreaks have been observed: (1) the presence of physical stressors, (2) boredom, (3) perceived pressures toward increased production, (4) strained labor-management relations, and (5) lack of interpersonal communication between workers. Clearly these factors can be considered contributors to stress and may thereby be linked to physiological reactions such as secretion of cortisol and catecholamines, and their consequences—increased blood pressure, rapid heart rate, and the like. More importantly, however, stress can also induce greater perception of bothersome or uncomfortable symptoms (e.g., Baum, Aiello, & Davis, 1979; Pennebaker, 1979). As Engel (1962) has noted, "What is felt includes not only such vague to precise mental content (the cognitive part), but also the awareness of physiological changes, such as palpitation, sweating, flushing, muscle tension or 'butterflies' in the stomach." These phenomena will beg explanation and the process by which workers go about labeling these internal sensations may have implications for MPI.

SOURCES OF STRESS IN OCCUPATIONAL SETTINGS

Physical stressors such as noise (Colligan, 1978; Shepard & Kroes, 1975; Stahl & Lebedun, 1974), air contamination (Shepard & Kroes, 1975) and noxious odors (Colligan, 1978; Colligan, Urtes, Wisseman, Rosenstell, & Anania, 1978; Folland, 1975; Kerckhoff & Back, 1968; Phillips, 1974; Shepard & Kroes, 1975) have been observed as antecedents of outbreaks of MPI. Glass and Singer (1972) and others have shown that unpredictable and uncontrollable environmental stressors can have behavioral costs and physiological consequences for people even after the cessation of the physical stressor. Stress has been linked to discomfort, illness, and a range of other problems (e.g., Frankenhaeuser, 1975). A factor that compounds the effects of physical stressors in the workplace is that most of the labor performed by workers involved in MPI outbreaks is assembly line work consisting of monotonous repetitive tasks. In their study of sawmill workers, Frankenhaeuser and Gardell (1976) found that people who did machine-paced short-cycle tasks exhibited stress reactions as measured by increased ratings of strain and ill health as well as by heightened catecholamine production. As Frankenhaeuser and Gardell (1976) note:

> The quest for increased productivity in industrial countries has been most intense in industrial manufacturing. This has enhanced the pressure on industrial workers as expressed in ever-increasing demands on pace and effective use of working time, and in decreasing options for variety, relaxation, and social interaction at work. To some extent the growing demands on the workers emanating from this development tend to be reflected in more frequent dispensary visits, sick leaves, and early retirements [p. 43].

Thus, stressful work environments can have health outcomes and those characterized by low levels of responsibility, variation, and satisfaction have been associated with heart disease and mortality (e.g., Caplan, Cobb, French, Harrison, & Pinneau, 1975).

It is interesting to note that in commenting on stressful environments, Frankenhaeuser and Gardell list three of the five common characteristics of settings in which MPI occurs. Boredom, increased pressure towards production, and limited contact with coworkers are indicated as stressors with health-related consequences. It should not be surprising, then, that environments in which these problems are confounded by environmental stressors and strained relations with management should be characterized by increased perception of illness symptoms.

This allows us to conclude that workers in settings in which MPI has occurred have probably experienced physiological symptoms of stress. Exposure to environmental stressors, boring repetitive tasks, and increased pressure all can predispose an individual to experience headaches, dizziness, nausea, fatigue, and the like, many of which are the same as symptoms found in most cases of MPI (e.g., Kerckhoff & Back, 1968). It is our intention to demonstrate that the ways in which workers attribute these symptoms will be of great importance in predisposing them to psychogenic illness and hysteria.

SOCIAL COMPARISON: AFFILIATION AND EMOTION

Social comparison theory as originally formulated (Festinger, 1954) suggests that people tend to evaluate themselves, that is, their beliefs and their abilities, through comparison with both social and nonsocial standards. In the absence of objective, i.e., nonsocial, standards, individuals will seek to compare themselves with others who are similar to them. In some areas, it is relatively easy for individuals to evaluate themselves by comparing their performance with some objective criterion. In other areas, such as attitudes, there is a significant absence of objective standards for comparison, and social forms of comparison become more important to the individual for both evaluation and validation of his or her beliefs. In such cases, we would expect a greater amount of comparison with other individuals. Although the general case for social comparison is usually made in terms of attitudes, the process seems equally applicable to emotions (cf. Schachter & Singer, 1962). Reactions to stress or anxiety may be especially difficult for an individual to evaluate objectively. More recent work extending the theory of social comparison still hinges on this basic mechanism (cf. Latané 1966; and Suls & Miller, 1977, passim).

Studies of affiliation indicate that some individuals do seek the company of others when they are in an anxious state. Schachter (1959) brought subjects into a laboratory where they were told the experiment was investigating the effects of electric shock. In the high fear condition, the shock was described in ominous terms: It would be intense and painful but would cause no permanent damage. Subjects in the low fear condition were told that they would receive very mild shocks that would not be the least painful. Subjects were then told that there would be a delay while the experimenter set up the equipment and were given a choice of waiting in a comfortable room with magazines and easy chairs or waiting in an empty classroom with other subjects. They were also asked to indicate how strong their preference was.

The results of this study indicated that the degree of anxiety was an important determinant of whether or not subjects chose to affiliate. In the high fear condition, 62.5% of the subjects chose to wait with others as opposed to only 33% of the subjects in the low fear condition. The preference to wait with others was also much stronger among highly fearful subjects.

Social comparison theory hypothesizes that individuals prefer to compare themselves with similar others. This notion is supported by another affiliation study (Schachter, 1959). Using a similar experimental paradigm, all subjects were given high fear instructions, but only half were given the choice of waiting alone or with other people waiting to be in the same experiment, that is, with people whose plight was similar to their own. The remaining subjects were asked to choose between waiting alone or with students who were waiting to see their advisors, that is, with people who presumably were in a different set of circumstances. Significantly more of the subjects who could wait with similar others chose to affiliate. Fearful individuals did not affiliate indiscriminately but tended to affiliate with others in a similar situation. Additional studies established that this selective affiliation was used by the subjects to evaluate the relative degree of their own arousal and its appropriateness for their situation.

Schachter also found large individual differences in the desire to affiliate that were related to birth order. First born individuals were most likely to seek out the company of others when fearful or anxious than were second borns, second borns were more likely to affiliate than third borns and so on. It might prove interesting to look at the birth order of individuals affected in an MPI outbreak, both as initiators or as followers.

In these studies, subjects were aware of the reason for their anxiety: The instructions were quite explicit about the nature of the shocks. The source of arousal outside of the laboratory may not always be so immediately recognizable. In such cases, individuals would search their environments for an explanation of their arousal.

To investigate the role of social comparison processes in evaluating physiological arousal, Schachter and Singer (1962) injected subjects with epinephrine, a drug which has sympathomimetic qualities. Some of these subjects (the Informed group) were given an objective referent for their arousal: The effects of the drug were fully explained to them. The remaining subjects (some uninformed as to the drug's effects, others misinformed as to its true effects) were told that the injection was a simple vitamin and were not led to expect the side effects actually produced, such as palpitation and tremor. All subjects were then placed in a room with an experimental confederate who played one of two roles in this situation: Within the experimental context, he became either increasingly euphoric or he became angrier and angrier as the experimental session progressed.

The results of this study indicate that social comparison has a strong influence on the interpretation of physiological arousal. Informed subjects were relatively unaffected by the confederate's behavior. In contrast, the other subjects—uninformed or misinformed—interpreted their arousal in terms of situational factors, i.e., the behavior of the other person. Uninformed subjects placed with a euphoric confederate rated themselves as being in a euphoric state while those with an angry confederate reported feeling a significant amount of anger. Behavioral observations also paralleled these findings. In addition, all comparisons were made to placebo injected subjects as well.

These results seem to indicate that two factors are involved in the identification of emotion: (1) physiological arousal and (2) cognitive labeling of that arousal. Labeling occurs as a function of establishing a cause and effect relationship between environmental stimuli and perceived arousal. New or ambiguous physical states, such as those associated with stress, will create pressure for people experiencing them to determine what it is they feel and what is responsible for their discomfort. It is possible, then, that MPI involves the perception of physical symptoms (induced by the stress characteristic of affected settings), and the way in which workers explain them. As Kerckhoff and Back (1968) point out, "To the extent that the new label 'makes sense' and is thus easily accepted, it will be adopted readily by all those who have the experience but have been unable to conceptualize it satisfactorily."

SOCIAL COMPARISON AND MPI

General Relationships

What we have proposed so far is that (1) work settings in which MPI has occurred are characterized by multiple stressors, (2) these stressors are responsible for changes in workers' physiological states, and (3) that many of these symptoms are the same as those reported in MPI outbreaks. We have also argued that workers will seek to understand their feelings, and that a primary source of information about them is other people. Thus, stressed workers will engage in social comparison in order to label their feelings. Under normal circumstances, workers would discuss these conditions and their reactions to them. By comparing their symptoms and their interpretations of them, these workers could be expected to come up with a reasonably accurate appraisal of stressful aspects of their jobs and their reactions to this.

In comparing epidemic hysteria with panics and riots, Sirois (1974) has noted that the conflictual object in the former is much less obvious. The "conflict object" is often unknown or not dealt with directly and it is frequently difficult or impossible for workers to openly acknowledge and deal with the conflict. Colligan and Murphy (1979) have noted little informational support prior to the outbreak of MPI and Kerckhoff and Back (1968) observed a greater reluctance among affected individuals to discuss job-related problems and work dissatisfaction. Smith, Colligan and Hurrell (1978) report that affected workers seem to be more financially dependent on their jobs than nonaffected workers (e.g., were more dependent on their jobs for total family income, had more children at home, would be more likely to quit if they didn't need the income, and so on).

In industrial settings that have experienced an outbreak of MPI, opportunities for communication between workers are generally limited. This limitation may be due to such factors as scheduling of lunch and break times (Stahl & Lebedun, 1974) or noisy environments (Colligan, 1978; Folland, 1975; Kerckhoff & Back, 1968; Phillips, 1974; Shepard & Kroes, 1975). Such a curtailment of communication would make it difficult for workers to engage in the open communication and discussion that might enable them to attribute their somatic complaints to the appropriate job-related stress stimuli. And in addition to limited communication among workers, communication between workers and management is frequently inadequate or misleading (e.g., Colligan, Urtes, Wisseman, Rosenstell, & Anania, 1978).

MPI Outbreak

In an unpleasant work situation containing stressed individuals who have limited opportunities for either objective evaluation or interaction with one another, there are two additional factors that are important in determining the outbreak of MPI;

a triggering event and an index case or cases. Most of the triggers in industrial settings appear to be strange odors while triggers in nonindustrial settings range from rumors to spirits. The lack of a specified triggering event is much more common in nonindustrial settings.

Although a precipitating event is not always necessary, it is generally present in industrial settings. An index case is obviously essential and provides a potential explanation for the worker's physiological reaction to the stressful environment. As stated by Colligan et al. (1978), "the detection of a strange odor . . . or witnessing others become ill may have provided the affected workers with an explanation of their discomfort and triggered a contagion reaction."

If social comparison processes do play an important role in the spread of MPI, some relationship between social groupings and contagion would be expected. Evidence along this dimension is presented by Kerckhoff, Back, and Miller (1965) and Stahl and Lebedun (1974).

In their analysis of the "June Bug" incident, Kerckhoff et al. noted a number of similarities among the affected workers: almost all were women, white, worked the first shirt, and were located in one area of the plant. Two months after the incident, affected and nonaffected women were asked to name their three best friends; the experimenters also had a record of the date on which each of the affected women had been "bitten." The results indicated more links between affected women and other affected women than between affected women and controls (unaffected women):

> Affecteds are twice as likely to be chosen solely by other affecteds, and controls are twice as likely to be chosen solely by other controls. Finally, of the mutual choices participated in by affecteds, three-fifths are with other affecteds. Similarly, of the mutual choices participated in by controls, two-thirds are with other controls [pp. 9–10].

The investigators divided the incidents of MPI into three major periods: "before," "Day 1," and "Day 2." Most of the cases (76%) occurred on Days 1 and 2, whereas 17% occurred beforehand. Of those who initially exhibited symptoms, 50% were social isolates. Although isolates appeared during other periods (at least 25% of the affecteds), they were most heavily concentrated in the initial phase before the mass outbreak.

In a study of a different outbreak which replicated and extended the June Bug study, Stahl and Lebedun (1974) interviewed 20 employees at a university data-processing center following an outbreak of MPI among 35 of the 60 workers employed there. An index of job satisfaction, a four point scale of symptom severity, and a sociogram of friendship patterns were constructed. Respondents indicated that the first person to become sick was a central figure in the sociogram analysis. Stahl and Lebedun noted that "many of the workers appeared quite aware of the temporal sequence of who got sick first and what the specific symptomatology was. This was especially true among workers who were involved in friendship networks with other workers. Social isolates, on the other hand, reported few symptoms [p. 47].'' Social networks also seemed to influence symptom severity. The friends of the workers with the highest severity scores (2 or 3) reported a mean severity score of 2.06, whereas the friends of those workers with lower severity scores (0 or 1) reported a mean score of 1.6. This would indicate that friends tend to exhibit similar symptomatology.

Interestingly, those individuals with the lowest level of job satisfaction tended to select as friends those workers who had the highest level of job satisfaction and vice versa. Although the Data Center was particularly limited in opportunities for communication between coworkers because of excessive noise and scheduling of lunch and break times, employees were able to form social networks (as evidenced by one respondent's complaint about the ''set group of ten girls who don't let anyone in the group.''). Given the limited amount of communication, workers may simply not have discussed job satisfaction with one another. Alternatively, if they did discuss this, the observed discrepancies in job satisfaction might have served to intensify the need to evaluate their reactions to their job rather than to validate them.

Both of the previous studies indicate that MPI spreads throughout a social network and provide support for a social comparison model. The Kerckhoff, Back, and Miller (1965) study shows that affected women tend to be friends with other affected women, while controls are friendlier with other controls, whereas the Stahl and Lebedun study demonstrates a relationship between friendship patterns and symptom severity.

These two studies provide an interesting contrast between an incident of MPI in which the initial cases were social isolates and one in which the index case was a socially central individual. Because individuals are most likely to compare themselves with their immediate reference groups, we would expect the spread of MPI to be much more rapid in those situations in which the initial case is centrally located in a social network. This appears to be the case: While the June Bug epidemic lasted about one week and affected 62 of 965 workers, the Data Center incident took place on two consecutive mornings and affected 35 of 60 workers.

The existence of social networks or the establishment of social groups within the work environment is not always a necessary antecedent to an outbreak of MPI; in fact, work environments in which MPI occurs tend to inhibit the development of social ties, by making it difficult for workers to communicate with each other. Although verbal communication may be difficult or impossible, the work setting generally is such that employees have considerable visual access to one another. When employees work together in a large open room, the ''attacks'' that an affected worker suffers can be visible to her coworkers. In the case of the more extreme symptomatology such as screaming and struggling violently with others (e.g., Chew, Phoon, & Mae-Lim, 1976), these attacks can scarcely be ignored. Colligan et al. (1978) found that nonaffected workers reported *hearing* of more people becoming ill than affecteds, while affecteds reported *witnessing* more people becoming ill (author's emphasis). Observing coworkers having attacks which they attribute to some environmental stimuli can certainly be a potent standard for comparison with one's own somatic complaints.

SOCIAL COMPARISON AND WORKER CHARACTERISTICS

In any incident of MPI, the total working population is not affected. Although affected workers do not differ markedly from nonaffected workers, several factors seem to be related to susceptibility to MPI. Of these, absenteeism, age and educational level, and ethnic background are ones that are relevant to a social comparison analysis.

Absenteeism

Prior work histories indicate a higher absenteeism rate among affecteds prior to the actual outbreak of MPI (Colligan & Murphy, 1979). Higher absenteeism means that these individuals will have even less communication with their coworkers, further decreasing already limited opportunities for social comparison. This, in turn, may inhibit labeling of job-related stressors and somatic complaints.

One possible explanation for the higher absenteeism among affecteds is that these workers are experiencing a stronger physical reaction to stressors in the work situation. Smith, Colligan, and Hurrell (1978) have noted that affected workers are more likely to report being bothered by environmental factors such as noise, poor lighting, irritating odors, and temperature variations in the workplace and indicate higher levels of job-related stress as a result of such factors as work pace, task repetition, job insecurity, and lack of supervisory support. Although the affected workers surveyed by Smith et al. (1978) reported poorer general health than unaffecteds, some studies using more objective measures of general health (i.e., previous health complications, reported medicine usage) have found no differences (Colligan et al., 1978; Colligan, 1978). Kerckhoff and Back (1968) found that affecteds were more likely to have been taking pills or other medication prior to the outbreak but were not more likely to have "had to stay home because of organic sickness only during the past year."

Alternatively, it is possible that these individuals are simply more prone to symptom reporting. Research has indicated that affecteds have a greater tendency towards somatization (Knight, Friedman, & Sulanti, 1965), score higher on the Hysteria scale of the MMPI (Colligan et al., 1978; Colligan, 1978), and are more likely to seek medical attention (Kerckhoff & Back, 1968). In a follow-up study conducted 25 years after an outbreak of epidemic hysteria in Tristan de Cunha, Rawnsley and Loudon (1964) found that affected women were later characterized by frequent medical consultations.

If the differential absenteeism rate is a function of greater symptom reporting or help-seeking, this could reflect a stronger need to evaluate physical reactions among affected individuals. This notion is supported by Kerckhoff and Back (1968) who found that affecteds tended to focus on emotive or physiological responses rather than potential external sources. If this is the case, such individuals should engage more actively in social comparison in an attempt to evaluate their physical distress and would thus be more susceptible to MPI.

Demographic Variables

Affected workers tend to be young individuals with lower educational attainment (Colligan & Murphy, 1979). Although this is to some degree true of the entire work populations in the situations where MPI has occurred, this limited experience and educational background should make these workers more susceptible to MPI. Older and better-educated people may have more knowledge of, or access to, objective referents that would allow them to accurately evaluate their somatic complaints and reduce pressures for social comparison.

Racial and Ethnic Background

Ethnic background is also an interesting variable, although susceptibility to MPI does not seem to be a function of race per se. Because individuals are most likely to compare themselves to similar others, social comparison theory would predict that an outbreak of MPI in an ethnically diverse setting would tend to be concentrated in one ethnic group. Support for this hypothesis is provided by the work of Chew and his associates (Chew, 1978; Chew, Phoon, & Mae-Lim, 1976). Chew discusses an incident that occurred in a large television assembly factory that employed 65% Malays, 23% Chinese, and 9% Indians. Of the 84 affected workers, 83 were Malays; no Chinese workers were affected and one Indian Muslim was only "frightened." In discussing an incident that occurred in a southern manufacturing plant, Kerckhoff, Back, and Miller (1965) note that "almost all of the victims . . . were white." Although data on the racial composition of the work force were not presented, Kerckhoff and Back (1968) did use black employees as a comparison group for affected workers.

Birth Order

Although information on birth order of affected workers is not available, such data could provide interesting insight into the role of social comparison processes in MPI. The finding that first borns are more likely to affiliate in a stressful situation (Schachter, 1959) was based on the presumption that first borns rely more heavily on social sources of evaluation and suggests that these individuals should be overrepresented among affected workers. As communication between coworkers is minimal prior to the outbreak first borns would be more likely to compare their symptoms with those of the initial cases and label them accordingly. In those cases where communication exists, or where first borns, under heavy evaluative pressure, initiate communication, the discussion of the circumstances that produced the stress-related symptoms may provide a nonpsychogenic explanation for the internal state and serve to circumvent or prevent the MPI outbreak.

WOMEN AND MPI

The most common characteristic of affected workers is their sex. In their survey of 16 incidents of MPI, Colligan and Murphy (1979) note that 93% of the workers reporting symptoms were women. Although the populations observed were often predominantly female, women were still disproportionately represented among the affecteds. This is apparently not just a modern-day phenomenon. Sirois (1974) notes that demonopathic manifestations from the 15th–19th centuries were almost exclusively female. In his historical survey, Markush (1973) reports many predominantly female incidents of epidemic hysteria (e.g., biting nuns in the fifteenth century, nuns mewing like cats in the 18th) and mentions that Hippocrates believed that hysteria was a result of "the wanderings of the uterus thirsting for semen." In 1883, Hirsch (cited in Markush, 1973) noted that women were affected either "exclusively" or "especially" by epidemic hysteria.

Although purely sociocultural explanations for the preponderance of women among affected workers are certainly possible (i.e., the female sex role making displays of illness more acceptable or leading to greater attentiveness and responsiveness to physiological states), we would like to take an approach that explores the interaction between biological and social factors.

One of the indisputable differences between males and females is that women have menstrual cycles. It is not our intent to debate the issue of whether menstrual symptoms (premenstrual syndrome, dysmenorrhea) are genuinely physiological reactions or the result of culturally transmitted beliefs and attitudes. Whether physiological or psychogenic in origin, physical and emotional discomfort immediately preceding and/or during menstruation are reported by significant numbers of women. Consideration of the menstrual effects suggests several alternative explanations for their role in MPI.

(1) It could be argued that the combination of a hormonal system already out of balance from the adrenal response to stress and the usual menstrual changes produces, in general, the sort of erratic behavior which folklore often ascribes to

women during the menstrual cycle. The tenability of this explanation, however, is minimal: Major reviews of the literature on menstruation fail to show any systematic erratic behaviors or psychopathologies associated with that condition (Ruble, 1977; Sommer, 1978).

(2) Recent psychological theory has emphasized attributional processes in shaping and determining behavior (Jones, Kanouse, Kelley, Nisbett, Valins, & Weiner, 1972; Shaver, 1975). This view would suggest that women, during menstruation, would attribute any unusual or untoward symptoms occurring at that time to the most salient cause—in this case, their menstrual period. Stress produced symptoms stemming from job-related conditions would be explained as but another manifestation of menstrual distress. Unfortunately for this view, the implication is that a definite fraction of the women workers would, therefore, be less susceptible to MPI (instigation) which would result in an overrepresentation of men among the MPI affecteds. This is patently at odds with the data.

(3) Menstruation may affect MPI by serving as an ancillary physiological factor: Social comparison may be heightened if there are internal states other than the stress induced ones discussed earlier, which also call for explanation. For example, if there were cases of incipient flu present, the general loginess and aches preceding the major outbreak might represent an internal state needing evaluation and could trigger off social comparison and its products—an homogenization of self-description and sick-like behaviors. The medical exams often given to MPI victims make it unlikely that flu and its symptoms are a feasible progenitor of the phenomenon, but other alternatives are worth considering.

As detailed earlier, MPI is more frequently observed in women than in men. It is possible that a uniquely feminine event, the menstrual cycle, is involved in MPI and contributes internal symptoms as well as providing a change in internal state, not classified as illness, which increases evaluative pressures. Of course, this possibility would only apply if a sizable proportion of the female workers were in synchrony with respect to their menstrual cycles. Although this is unlikely on a purely chance basis, there exists some evidence that such synchrony occurs as a consequence of female participation in the same work group.

McClintock (1971) has briefly reviewed the animal literature on synchrony of oestrous and concluded that conjoint living produces such synchrony in mice and other species. More cogently, she conducted a study of college students in a dormitory and demonstrated that over the course of the school year, both close friends and roommates, although asynchronous at the start of the year, had closely correlated cycles just a few months later. Appropriately analyzed control groups rule out the possibility that the school pressures, exams and the like produced synchrony. Equally germane to this discussion is the report by McClintock that 7 female lifeguards, whose menstrual onsets were scattered at the beginning of the summer, after 3 months of working together, had cycles with onsets within a 4-day span. This suggests that synchronous work cycles and physical proximity may be sufficient conditions to trigger menstrual synchrony.

There are indications, but weaker ones, that similar coordination of endocrine functions exists for a variety of other cyclically produced hormones as well (e.g., Aschoff, 1976; Fulkerson, 1978; Kaul & Berdanier, 1975; Orth & Island, 1969; Zucker, 1976). These cycles may also provide pressures for evaluation. Although we are not attributing MPI to menstrual distress we are suggesting that the evaluative activities triggered off by work related stressors might well be augmented by a variety of synchronous endocrine changes, of which menstruation is the best known. These changes simultaneously affect a large part of the work cohort with yet another internal state to evaluate. The menstrual history of women involved in previous MPI outbreaks has not been reported, but might be worth monitoring in future occurrences.

CONCLUSION

There are two major problems in understanding mass psychogenic illness in industrial settings. The first is explaining why it occurs; the second is explaining why it does not always occur. Logically, there is no reason why MPI could not be a similar product of several different causes, but the consistencies in the characteristics of workers, settings, and events noted both by previous reviews and in this paper suggest that a similar process is at work. Of the many possible processes which could be utilized as an after-the-fact explanation, we have made a case for social comparison and we have done so for several reasons:

1. The theory of social comparison processes provides a rather neat conceptual fit to the data. It can explain a way in which MPI comes about.
2. The theory can explain the circumstances under which MPI does not occur, that is, it can point to conditions likely, other things being equal, to prevent MPI.
3. The theory can help specify some of the characteristics of those people likely to be susceptible or resistant in a particular case of MPI.
4. The theory can point to particular measurements, e.g., birth order, menstrual data, which can provide inferential support for intermediate processes of the proposed mechanism. It has the virtue of being falsifiable both in whole or part, and, hence, modifiable.
5. The explanation provided by the theory is concrete enough to provide a guide for experimental tests of MPI either by analogy in the laboratory or directly in the field.

One final word. MPI, as with many other psychosomatic illnesses, is a real illness. That there is economic loss is clear, but it should also be noted that the victims of it are ill—the lack of a specific pathogen to be held accountable does not mitigate their real distress. It would only compound their illness to imply that because the instigating factors are psychological their illness is "only in their heads." They would then be justified in voicing remonstration similar to the one contained in the old textile song, *Hard Times in the Mill,* "Ain't it enough to break your heart, You work all day and at night it's dark."

REFERENCES

Aschoff, J. Circadian systems in man and their implications. *Hospital Practice,* 1976, *11*(5), 51-57.

Baum, A., Aiello, J., & Davis, G. *Neighborhood determinants of stress and symptom perception.* Paper presented at the Meeting of the American Psychological Association, N.Y.: September 1979.

Caplan, R., Cobb, S., French, J. R. P., Jr., Van Harrison, R., & Pinneau, S. R., Jr. *Job demands and worker health,* HEW Publication. National Institute for Occupational Safety and Health (NIOSH), Washington, D.C., 1975.

Chew, P. K. How to handle hysterical factory workers. *Occupational Health and Safety,* 1978, *47*(2), 50-53.

Chew, P. K., Phoon, W. H., & Mae-Lim, H. A. Epidemic hysteria among some factory workers in Singapore. *Singapore Medical Journal,* 1976, *17*(1), 10-15.

Colligan, M. J. *An investigation of apparent mass psychogenic illness in a furniture assembly plant.* Unpublished NIOSH evaluation report, 1978.

Colligan, M. J., & Murphy, L. R. Mass psychogenic illness in organizations: An overview. *Journal of Occupational Psychology,* 1979, *52,* 77-90.

Colligan, M. J., Urtes, M. A., Wisseman, C., Rosenstell, R. E., & Anania, T. L. *An investigation of apparent mass psychogenic illness in an electronics plant.* Unpublished NIOSH evaluation report, 1978.

Engel, G. L. *Psychological development in health and disease.* Philadelphia: Saunders, 1962.

Festinger, L. A theory of social comparison processes. *Human Relations,* 1954, *7,* 117-140.

Folland, D. S. *Suspect toluene exposure at a boot factory.* Internal report, Tennessee Department of Health, Nashville, Tennessee, 1975.

Frankenhaeuser, M. Experimental approaches to the study of catecholamines and emotions. In L. Levi (Ed.), *Emotions: Their parameters and measurement.* New York: Raven Pres, 1975.

Frankenhaeuser, M., & Gardell, B. Underload and overload in working life: Outline of a multidisciplinary approach. *Journal of Human Stress,* 1976, *2*(3), 35-46.

Fulkerson, W. J. Synchronous episodic release of cortisol in the sheep. *Journal of Endocrinology,* 1978, *79,* 131-132.

Glass, D. C., & Singer, J. E. *Urban stress.* New York: Academic Press, 1972.

Jones, E. E., Kanouse, D. E., Kelley, H. H., Nisbett, R. E., Valins, S., & Weiner, B. *Attribution: Perceiving the causes of behavior.* Morristown, N.J.: General Learning Press, 1972.

Kaul, L., & Berdanier, C. D. Effect of meal-feeding on the daily variations of insulin, glucose and NADP-linked dehydrogenases in rats. *Journal of Nutrition,* 1975, *105,* 1132-1140.

Kerckhoff, A. C., Back, K. W., & Miller, N. Sociometric patterns in hysterical contagion. *Sociometry,* 1965, *28,* 2-15.

Kerckhoff, A. C., & Back, K. W. *The June Bug: A study of hysterical contagion.* New York: Appleton-Century-Crofts, 1968.

Knight, J. A., Friedman, T. S., & Sulanti, J. Epidemic hysteria: A field study. *American Journal of Public Health,* 1965, *55,* 858-865.

Latané, B. (Ed.) Studies in social comparison. Supplement 1 to *Journal of Experimental Social Psychology,* 1966.

McClintock, M. K. Menstrual synchrony and suppression. *Nature,* 1971, *229,* 244-245.

Markush, R. E. Mental epidemics: A review of the old to prepare for the new. *Public Health Reviews,* 1973, *2*(4), 353–442.

Orth, D. N., & Island, D. P. Light synchronization of the circadian rhythm in plasma cortisol (17-OHCS) concentration in man. *Journal of Clinical Endocrinology,* 1969, *29,* 479–486.

Pennebaker, J. *Environmental determinants of symptom perception.* Paper presented at the Meeting of the American Psychological Association, N.Y.: September 1979.

Phillips, P. E. *Internal report prepared for the Division of Health of the State of Missouri,* 1974.

Rawnsley, K., & Loudon, J. Epidemiology of mental disorder in a closed community. *British Journal of Psychiatry,* 1964, *110,* 830–839.

Ruble, D. N. Premenstrual symptoms: A reinterpretation of symptoms. *Science,* 1977, *197,* 291–292.

Schachter, S. *The psychology of affiliation.* Stanford: Stanford University Press, 1959.

Schachter, S., & Singer, J. E. Cognitive, social, and physiological determinants of emotional state. *Psychological Review,* 1962, *69,* 379–399.

Shaver, K. G. *An introduction to attribution processes.* Cambridge, Mass.: Winthrop, 1975.

Shepard, R. D., & Kroes, W. H. *Report of an investigation at the James plant.* Internal report prepared for the National Institute for Occupational Safety and Health: Cincinnati, Ohio, 1975.

Sirois, F. Epidemic hysteria. *Acta Psychiatrica Scandinavica,* Supplementum, 1974, *252,* 5–46.

Smith, J. J., Colligan, M. J., & Hurrel, J. J., Jr. Three incidents of industrial mass psychogenic illness. *Journal of Occupational Medicine,* 1978, *20*(6), 399–400.

Sommer, B. Stress and menstrual distress. *Journal of Human Stress,* 1978, *4*(3), 5–10, 41–47.

Stahl, S. M., & Lebedun, M. Mystery gas: An analysis of mass hysteria. *Journal of Health and Social Behavior,* 1974, *15,* 44–51.

Suls, J. M., & Miller, R. C. *Social comparison processes: Theoretical and empirical perspectives.* Washington, D.C.: Hemisphere Publishing Company, 1977.

Zucker, I. Light, behavior and biologic rhythms. *Hospital Practice,* 1976, *11*(10), 83–91.

10 Theories of Contagion As They Relate to Mass Psychogenic Illness

Jonathan L. Freedman
University of Toronto

ABSTRACT

This chapter presents a review of the theories regarding contagion. It discusses the original ideas of Le Bon and Freud that being in a crowd releases primitive impulses and the modern reworking of these ideas in terms of deindividuation. The chapter also deals with the theory that contagion is essentially imitation, mediated by restraint release due to observing another perform an action that the individual is in conflict about performing himself. The final major idea is that the amount of space available (density) as well as the number of people who are present affect the likelihood of contagion occurring. In addition, contagion is distinguished from phenomena such as conformity and the emergence of group norms. The various theories are criticized with particular reference to mass psychogenic illness. It is concluded that the theories may provide a framework for considering incidents of mass illness, but that none of them alone nor all of them together offer a full explanation of the phenomenon.

Mass illness due to psychological causes appears to be a classic instance of a phenomenon that has been called contagion. In 1895 LeBon suggested that in a mob all of the people tend to feel and behave the same way because the emotions and actions of one person spread through the group—a process he termed "contagion." Most of his observations were based on the actions of large groups of people engaged in violent or other antisocial behavior, though it should be noted that he also realized that more positive or prosocial behaviors might also be involved in contagion. He cited the mob's storming of the bastille during the

French revolution as one instance of contagion which he considered deplorable (though obviously others would view it in a different light).

Whether the actions were positive, negative or neutral, the critical element for Le Bon was that in a mob the people behaved in ways that they would not have if they were alone, and that by themselves each person might act differently, while in the group they all felt and acted the same. This is probably an overstatement of the situation. It is rare that every individual feels and acts identically just because they are in a group. However, the basic idea is sound. There are clearly times when people in a group engage in some activity, or in the present case, exhibit certain symptoms, that few if any would by themselves. However, in order for this to be an instance of contagion, it must follow a particular pattern. Namely, one or at most a small number of people in the group must first exhibit the behavior, and then this must spread through the group so that others also exhibit it.

It is important to distinguish this process from others that can also cause the members of a group to behave similarly. First, it may be that there are sufficient reasons for the people to behave in a particular way, and these reasons affect everyone at the same time. For example, if a fire breaks out in a factory and everyone reacts by becoming frightened, and running from the building, it would ordinarily not be considered contagion. Rather, everyone is merely reacting in more or less the same manner to a clearly discernible external event. More to the point, if an instance of mass illness can be shown to be due to a gas leak, bacterial contamination, or sudden overwhelming stress, we would not consider these examples of psychogenic illness nor of contagion. Thus, in investigating mass illness, the first focus is always on whether some physical cause exists, and the outbreak then has little psychological or sociological interest unless people have greatly overreacted to the situation.

A second process that may be more difficult to distinguish from contagion is usually called conformity. The individual does what other members of the group do because of social pressure. In the typical instance of conformity, everyone in a group acts in one way and a new member or someone who has not yet had the opportunity to perform the behavior, also acts that way. Note that in this case, most or all of the group is already performing the act, and it is this that puts pressure on others. In contrast, contagion occurs when only one person or at most a small percentage of the group perform the action and the rest go along with it. In other words, it is quite different from conformity in that in contagion a small number affect the majority, whereas in conformity, it is the majority that affects the rest. However, it is possible that conformity could occur when one especially outstanding or influential member of a group behaves in some manner, and the rest go along because of his or her position. That is, under some circumstances the norm might be established not by the majority, but by a small percentage of the group who happen to be particularly influential.

This leads to a third process that has been called emergent norms (Turner, 1964). According to this idea, there are times when everyone ends up behaving the same because the group has established a norm that a particular form of behavior is both acceptable and desirable. The group need not get together and have a vote on this. Rather, and more typically, the norm emerges gradually and each member identifies it for himself. This occurs because the members observe what others are doing, decide for themselves what seems to be appropriate under the circumstances, decide further that everyone agrees on what is appropriate, and then behave accordingly. If they decide that the group is storming the bastille, they storm the bastille; if they decide that the group is running into a burning building to save someone, they do that; and if they decide that the group is having headaches or laughing fits, they do that. In other words, these are instances of conformity to group norms. Each individual is exposed to these norms, is under pressure to go along with them, and thus we get similar behavior from the group members. Although this obviously occurs under some circumstances, it seems not to be applicable to most instances of mass illness. The notion that there is a norm to become seriously ill is highly implausible, and this theory does not explain how such a norm might develop in the first place. Surely, one or two people exhibiting physical symptoms would not ordinarily cause anyone to think that this was the norm. Thus, although emergent norm theory may account for the latter stages of contagion, when a large number of members are already exhibiting the behavior, it is not helpful in explaining why contagion occurs in the first place.

Let us be clear then that contagion occurs when, and only when, the behavior of one individual or at most a few individuals is observed by others, and these others then perform the behavior. Moreover, in order for it to be contagion, the following criteria must also be met:

1. there is no sufficient reason for the people to be behaving in this manner apart from the fact that others are;
2. closely related to the first, those who observe the others would be unlikely to perform the behavior by themselves—that is, watching the others behave increases the likelihood of the behavior occurring;
3. it is not merely a case of members of a group imitating or being otherwise influenced by an especially powerful or popular individual, though such factors may play a role (this is to distinguish contagion from instances where leadership is involved, such as when an influential member says ''Lets all go swimming'' or merely goes off to swim - and everyone follows.)

Finally, it is perhaps important to note that in most instances of contagion, maybe all, the behavior in question involves some conflict. The individual may have some reason to perform it, but also some reason not to—he is, in traditional psychological terminology, in an approach-avoidance conflict. For example, the Frenchman who storms the bastille probably has reasons for wanting to destroy this symbol of the government, but is also afraid of the consequences and may feel some hesitation due to moral restraints against breaking the law even though he hates the government that passed the law. Similarly, people who stand in approval at the end of a piano concert want to show how much they enjoyed the performance, but each person is also reluctant to be the first to stand from fear of embarrassment (if no one follows suit) or simply from not wanting to be too conspicuous. In the case of psychogenic mass illness, the reasons for displaying the symptoms are less obvious, but according to the case studies, there always seems to be some particular stress present, or other causes; although the pressures against displaying the behavior are clearer, involving social norms, needs of work, and so on.

Taking these as criteria for contagion, it seems that mass illness typically satisfies them quite well. First, in all of the instances that have been described in any detail, the illness appears in a small number of individuals, and then many others exhibit the symptoms. That is, the illness does not break out in a great many people simultaneously—rather it starts with a few and "spreads." Second, there is no known physical cause for the illness, nor are there any sufficient psychological causes such as stress. Often, of course, the individuals are under considerable stress, but they have been for some time before the outbreak and are again after the outbreak. Thus, the stress is not enough to cause the illness by itself—people can experience this level of stress (or other pressures) without showing the particular symptoms of the illness; and indeed, the very individuals who become "ill" have experienced identical conditions before without becoming sick. Third, as far as we can tell, the outbreaks are not organized by any person or persons, nor are they due to people following a particularly influential person who wants others to copy him or her. Instead, a few people get sick, they do not try to make others get sick also, and the others are not consciously imitating them. Those involved appear to believe that they are sick; they are not deliberately exhibiting the symptoms. There may be other ways of considering psychogenic illness, but clearly it is one of the most dramatic instances of contagion. Therefore, theoretical explanations of contagion may help us understand the causes of mass psychogenic illness.

The earliest explanation of contagion was that being in a group in some way released primitive impulses by suppressing the individual's normal restraints. Both Le Bon (1895) and in somewhat different terms Freud (1923) subscribed to this view. The idea was that the usual moral sense and system of controls that we each carry around with us depends on feeling to some extent responsible for our own actions. When we are no longer responsible for ourselves, when we can blame others for what we do, or when for some reason others take responsibility for our behavior, our control system is weakened or even breaks down entirely. The superego, which ordinarily specifies what actions are allowable, becomes irrelevant when we feel that behavior is either beyond our control (we are forced to do it) or when no one, including ourselves, could blame us for what we do (one cannot feel guilty if the action is not freely performed, or when someone says that they take full responsibility—it is their guilt, not ours). Being in a group makes us feel less responsible for our actions. Instead of acting as individuals, we act as a group, and the responsibility is shared. Moreover, the group comes to act as a unit, submerging each individual's freedom of action and moral sense in the "group mind." Thus, not only is the individual less responsible because of shared responsibility, but he is carried along by the group mind. Moreover, and perhaps most important, once the superego is submerged or weakened, the more primitive impulses of the *id* (in Freudian terms) are released. They have been held in check by the superego and are now free to be expressed more fully. Thus, aggressive, destructive and sexual impulses, that are usually carefully restrained, are more likely to be expressed in groups.

Although this is a fascinating idea, there is no research directly supporting it. The only work that is possibly relevant will be discussed below in the context of a slightly different theory. However, whether or not the explanation is correct, it would seem to offer us little regarding our current concerns. In most instances of mass illness, the individuals affected are not displaying especially primitive behavior. They are not acting aggressively, destructively, or sexually. If, for example, there were an outbreak of wanton destruction in a factory, that would fit the Freudian model; or if a group of female workers suddenly threw off their clothes and leapt on the available males or on each other, there would be some reason to consider the Freudian view relevant. Indeed, it has been suggested that the outbreaks sometimes seem to have sexual overtones, but far from being overt, the sexuality, if it is there, is generally disguised. Thus, it is difficult to see how the notion that being in the group releases primitive impulses or relaxes the hold of the superego would account for mass illnesses. One would have to assume that people unconsciously want to become ill or experience the particular symptoms that appear, and that they have held these desires in check, and now are expressing them because of the group effect. This seems highly implausible.

On the other hand, perhaps it could be argued that this is exactly what happens with one limitation—the impulses are being expressed more openly, but the superego still has some strength and thus the impulses, which usually are fully suppressed, appear in disguised form. Nevertheless, this viewpoint still leaves rather vague why the contagion itself occurs. That is, it gives a reason why people in groups may engage in behaviors they would not engage in by themselves, but it does not explain why, at some point, the behavior of one person spreads to the rest. Nor, for our purposes, does it explain why, after exposure to the same conditions for many months or years, the people suddenly exhibit the symptoms. Thus, this early view is helpful mainly as a description of people in groups but not as an explanation of the specific phenomenon with which we are concerned.

More recently, a number of psychologists have translated these ideas into slightly different terms. Leon Festinger used the term ''deindividuation'' to refer to the phenomenon in which an individual's personality and will is submerged in a group. According to this view, people in a group tend to lose their sense of personal identity. Instead of thinking of themselves as individuals, there is a tendency for them to think of themselves as group members. This causes them to identify closely with the group, with the other members of the group, and with the goals and action of the group. At the same time, because of this close identification plus the loss of the sense of their own identity, they feel less responsible for their own actions. Whatever the others do, they do, and if one person does something, each individual is likely to do the same merely because they are all members of the same group and because each person's control is weaker then it would be outside the group.

The emphasis in this view is not on the release of primitive impulses, but rather on the lowering of the normal restraint and instead of using psychoanalytic concepts, it discusses the phenomenon in the terminology of research psychologists. According to the deindividuation explanation, it is not that new impulses appear, but that various reasons for restricting behavior are weakened. The individual fears punishment less, because the blame is spread throughout the group; he feels less guilty, because in some sense others are causing him to act the way he did; he need not fear social ostracism or rejection, because he has the support of the group; and so on. All of this allows the person to act in ways he would not if alone.

The evidence for this view and the somewhat similar views of Le Bon and Freud comes from two kinds of research. The first concerns how being in a group affects the amount of aggression or antisocial behavior expressed. This work consists of comparing people's behavior when they are alone with when they are in a group. There is very little systematic work, and somewhat surprisingly, what there is runs exactly counter to what would be expected from the theory. Rather than being more aggressive or antisocial in groups, people tend to be less aggressive and antisocial. For example, one study (Diener, 1976) observed young adults in a situation in which they were free to act aggressively toward someone who had not harmed them in any way. In all conditions, the adults observed someone else acting aggressively toward the ''victim,'' thus providing a situation that allowed contagion to occur. However, adults who were alone with the victim expressed much more aggression than those who were in a group. In summarizing all of the research on this issue, Diener (1977) concludes that the evidence does not generally support the idea that group membership increases antisocial behavior. Sometimes it does, but there is no reason to believe that in general people act more ''primitively'' in groups than when they are alone.

In fact, both research and observation indicate that membership in a group often is an important factor producing compliance with social norms. Group pressure tends to make individuals act in accordance with the norms and rules of that group, and also generally to act in more socially-accepted ways than when they are alone. Thus, being in a group appears to have varied effects on individual behavior. Under most circumstances it causes people to behave more conservatively, more acceptably, and therefore less "primitively." In contrast, under some circumstances it may have the opposite effect. The crucial factor is that the individual will tend to do what he or she thinks the group thinks is correct. Therefore, if the group for some reason holds antisocial values (as in some delinquent groups), someone joining the group may behave more antisocially than he would alone; whereas if the group holds more socially acceptable values, the individual will behave in socially acceptable ways. The group does not appear to release any primitive impulses—rather it generally puts pressure on the individual to behave in certain ways determined by the group. Perhaps, under some unusual conditions, a normally prosocial group will become antisocial and exhibit an unusual amount of aggressive or destructive behavior; but this is not typical. Thus, the idea that groups necessarily cause individuals to behave in primitive ways seems to be unfounded, and this aspect of the theories is probably incorrect.

The second source of evidence, especially for the idea of deindividuation, concerns the effect of anonymity on behavior. The argument is that being in a group makes each person stand out less, causes him to be less easily identified, and this results in the persons' feeling deindividualized. At one level, this certainly makes sense intuitively. Certainly when we are in a crowd of 50 people or even in a group of five, we are to some extent considered part of the group rather than as individuals. Someone alone on the street, with no one near him, is more visible and more easily identified than someone walking in a crowd. He stands out more, his behavior is more noticeable—whatever he does, it is clearly the individual doing it. When a group does something, it is often difficult to know just who in the group participated. We know the group did it, but often cannot attribute the actions to any given individual. Similarly, in a factory or office, someone who is alone in a room or is the only one who has a particular job is more easily identifiable than someone in a room with others or who has a job similar to that held by many others. Thus, group membership does tend to reduce individual recognition. The second step in the argument is that this reduced recognition leads to the individual feeling less responsible for his own behavior, feeling deindividuated, and therefore freer to act in ways he would not by himself. Therefore, degree of anonymity becomes a crucial variable.

Quite a few studies have investigated the effect of anonymity on behavior. Although the research is not terribly convincing, the results do generally indicate that people who feel anonymous are more likely to engage in antisocial behavior. Several experiments showed that not mentioning people's names or actually hiding their faces had this effect (Festinger, Pepitone & Newcomb, 1952; Singer, Brush, & Lublin, 1965; Zimbardo, 1969). One study demonstrated that children who were trick-or-treating and wearing Halloween masks were less likely to steal extra candy when an adult first asked their names (Diener, Fraser, Beaman, & Kelem, 1976). And Watson (1973) reported that cultures in which warriors wore masks or otherwise disguised their faces were more likely to engage in especially aggressive acts during warfare than cultures in which warriors did not disguise themselves. On the other hand, at least one study (Diener, 1976) found that anonymity had no effect on aggressiveness.

In any case, there is a serious difficulty with this work. The original idea was that anonymity would lead to deindividuation, but none of these studies provides evidence for this. An easy alternative explanation of increased antisocial behavior is simply that anonymous people are much less likely to worry about actual consequences of their actions. Because they cannot be identified, they are unlikely to be punished for what they do. That is, no complex psychological explanation is required—the effect may be due to the realistic, practical consideration of whether you can get away with it. Thus, there is little or no evidence to support the existence of a state that might be called deindividuation, and certainly no evidence to show that it is involved in contagion.

The ''primitive expression'' and ''deindividuation'' theories are quite similar even though they use different terminology. Both view the individual in a group as feeling less responsible for his action, and to some extent ''caught up'' in the actions of others. His own built-in controls are weakened, and he is therefore more likely to do what others do, and especially more likely to engage in antisocial behavior and behavior that runs counter to his own moral and ethical sense. Although the evidence for this view is at best meager, it remains an interesting idea.

Assuming for the moment that being in a group does reduce one's sense of personal responsibility, and that this in turn leads to a tendency to do what others in the group are doing, and that it also allows some behavior that might otherwise not occur, the crucial question is what factors determine when all of this will happen. Clearly, not all groups produce these effects all the time. Returning to our concern, incidents of mass psychogenic illness are extremely rare. People work in groups a great deal of the time, they are exposed to stress much of the time, and an occasional person in these situations must exhibit dramatic psychogenic symptoms. That is, isolated cases of fainting, feeling sick, becoming nauseous and so on probably appear with some regularity in factories, offices, and other group settings. Yet, the contagion of these symptoms resulting in mass illness occurs in only a miniscule percentage of the cases. What are the crucial factors that trigger the spread of the symptoms? Even if being in the group has the effects described above—of reducing responsibility and increasing imitation within the group, we do not yet have an explanation of why sometimes, quite rarely, this situation leads to mass illness. Thus, this general theoretical approach, while it may provide some framework, offers no answers.

Another theory that may have some explanatory power is in terms of imitation. As proposed by Redl (1949) and Wheeler (1966) the idea is that contagion is essentially imitation that is mediated by the reduction of restraint caused by observing someone else's behavior. As noted previously, the individual is in a conflict situation—he has some reasons for behaving in a particular way, but other reasons for not behaving in that way. He may experience stress, feel somewhat angry, and want to throw a rock through a window; but he is afraid of being punished by the law, afraid of looking foolish, worried about acting antisocially, and so on. Under most circumstances, for most people, the restraints are stronger than the impulse and so the rock is not thrown. However, once he observes someone else throw a rock, the idea is that this weakens the restraints and then he is freer to perform the action.

Certainly, there is evidence that observing someone perform an action and not get punished for it, increases the likelihood that others will imitate the act; whereas if the first person is punished, this reduces the likelihood of others performing it (Bandura, Ross, & Ross, 1963). The act need not be especially antisocial to fit this model. For example, many people may be reluctant to be the first one to step onto a dance floor in a night club because they may be afraid of looking foolish or just standing out. Once someone else "takes the plunge" and starts dancing, others join in. Similarly, once someone has applauded at a performance, others tend to applaud; once someone offers help to someone in need, others are more likely to help, etc. The general point is that once you have observed someone else performing an action that you are somewhat inclined to perform yourself, but which you are for whatever reason reluctant to perform, your reluctance may be reduced and you would then be more likely to perform the action. Wheeler (1966) has provided some evidence that this occurs, and other research (Freedman, Birsky, & Cavoukian, 1980) also demonstrates the effect.

The question is whether this explains the incidents of mass illness. Clearly, at some level these are incidents that involve imitation. One person exhibits symptoms, often bizarre symptoms, and then others exhibit the identical or similar symptoms. It could be argued that what has occurred is that many people are inclined to show their displeasure at their situation, or to express their negative feelings due to stress, or in some way make a statement about their dislike of where they are or the conditions under which they work. They have probably not specifically considered throwing a fit, fainting, or any particular symptom. But they are tempted to do something. Against this are a variety of reasons for not doing anything along these lines because they will be disciplined by their superiors, embarrassed in front of their peers, or whatever. Once someone, for some reason, does exhibit the symptoms, these other people are less afraid of the negative consequences and they too want to express themselves. Although they would not have considered this particular form of expression, it is a convenient way of expressing themselves and they therefore imitate the initiator.

All of this makes sense and probably does play a role in mass illness. However, just as with the previous theory, it leaves unexplained why the illness breaks out at a particular time and not at others, why it occurs rarely rather than often, and also offers little in the way of suggestions for reducing the incidence of mass illness.

A final theoretical approach to contagion is quite different from the others. Rather than focus on the mechanisms involved, it concerns the conditions under which contagion is likely to occur. In particular, it is based on research on how the amount of space and the number of people present affect people's behavior. The basic idea is that increased density (i.e., less space available per person) tends to intensify people's reactions to each other. This so-called intensification theory (Freedman, 1975) states that as density increases, other people become a more important and salient aspect of the environment, the actions of others are more likely to be noticed and to impinge on the individual, and therefore reactions are magnified. An analogy might be our reaction to music. If we like the music, as it is played louder, we notice it more, it becomes a more important stimulus, and we like it more; in contrast, if we dislike the music, as it becomes louder, we dislike it more. Similarly, our reactions to other people become stronger as the density increases. Whatever our reactions under low density, they are stronger under high density. If we respond positively under low density, we respond more positive under high density; if we respond negative under low density, we respond more negatively under high density. A considerable amount of research has supported this theory.

This intensification effect has been shown also to apply to contagion. The idea is that contagion depends in part on observing what someone else is doing, on being affected by it, and the more important the other people are, the more contagion should occur. That is, just as other reactions to the people are intensified by high density, so the likelihood of contagion is magnified. This has been supported in two studies—one showing that the contagion of laughter was increased by high density (Freedman & Perlick, 1979), the other that the contagion of applause was increased by high density and also, to a lesser extent, by larger groups (Freedman et al., 1980).

The application of this theory to mass illness would be that we would expect the possibility of contagion to be greater when people are under relatively high density, and also when they are in larger groups. Thus, working conditions that provide little space per person, and in which many people work, should be more likely to lead to incidents of contagion. If we assume that mass psychogenic illness is due to contagion, it should occur more under these conditions.

It should be clear from this review that none of the theories of contagion provides a full account of the phenomenon, and certainly does not explain in detail why mass psychogenic illness occurs. The various theories give us some hints as to why contagion occurs at all, and what factors may increase or decrease the chance of contagion. Taking all of the theories at face value and accepting them for purposes of the argument, would lead to the following hypotheses:

1. Mass illness should occur more when individuals feel part of a group, and when they feel relatively anonymous. Situations in which people are all working at similar jobs, when they are treated with relatively little individual attention, when they are "cogs" in a machine or part of a large institution, should all increase contagion. Moreover, people who are in this situation should be more likely to be involved in mass illness than people who are more differentiated in terms of jobs, special attention, or whatever. It would follow from this that anything that increases people's sense of being individuals, that causes them to think of themselves as separate from the rest of the group in some way (not necessary opposed to the group, but merely not just one of many), should reduce the chances of mass illness occurring.

2. Assuming that contagion is due in part to imitation mediated by restraint reduction, the response to the first instance of illness should play a role in the contagion. When someone exhibits symptoms, the idea is that this reduces the restraint others may feel and therefore increases the chance that they will show similar symptoms. However, this assumes that the first person showing the symptoms suffers no negative consequences. If, instead, the first person were disciplined, treated harshly, rejected socially, or given negative feedback of any kind, this would presumably reduce the likelihood of further contagion. Of course, this is usually not a practical or ethical possibility. When someone is sick, they are and should be given good treatment, special consideration and so on. This humane treatment, which is certainly appropriate, increases the likelihood that the illness will spread by contagion. It may be possible to devise good treatment that is nevertheless not reinforcing. For example, the "sick" individual could be immediately isolated from the others, thus reducing the chance of contagion and perhaps also implying some negative consequences. Whether this is practical or would work is questionable, but at least a range of responses should be considered.

3. High density, especially when combined with large groups, should increase contagion and the likelihood of mass illness. It should be clear that high density working conditions are not necessarily bad—the evidence is that high density does not generally have negative effects on people. High density may lead to more positive group feelings and higher productivity under some circumstances. Indeed, the stronger group feelings induced by high density are generally considered a positive outcome, though in the present context they would also increase the possibility of mass illness. On the other hand, it does seem that lowering densities and reducing the size of groups should reduce the chance of mass illness by reducing contagion. However, this change is not necessarily a good one overall, considering that it may also reduce positive group feelings. Perhaps an optimal level of density and group size could be found for each environment such that it was high enough to encourage positive social interactions, but low enough to keep contagion to a minimum.

Considering all of the theories, and the factors just mentioned, we still do not understand why mass illness occurs at a particular time when similar conditions have existed previously. The theories provide some framework within which to consider the phenomenon and offer some suggestions as to factors that affect the likelihood of mass illness occurring. But chance elements almost certainly play a role, as do various factors that have not yet been specified. It is important to keep in mind that mass illness is a very rare event. The theories specify factors that increase contagion and therefore increase the likelihood of mass illness; but mass illness will not occur often no matter what the circumstances. Thus, these factors should be seen only to increase the chance slightly, perhaps in the way that in golf, holding the club just right, and having a perfect swing increases the chance of a hole in one, but holes-in-one are highly unlikely events regardless of what one does.

REFERENCES

Bandura, A., Ross, D., & Ross, S. A. Vicarious reinforcement and imitative learning. *Journal of Abnormal and Social Psychology,* 1963, *67,* 601–607.

Diener, E. Effects of prior destructive behavior, anonymity, and group presence on deindividuation and aggression. *Journal of Personality and Social Psychology,* 1976, *33,* 497–507.

Diener, E. Deindividuation: Causes and consequences. *Social Behavior and Personality,* 1977, *5,* 143–155.

Diener, E., Fraser, S. C., Beaman, A. L., & Kelem, R. T. Effects of deindividuation variables on stealing among Halloween trick-or-treaters. *Journal of Personality and Social Psychology,* 1976, *33,* 178–183.

Festinger, L., Pepitone, A., & Newcomb, T. Some consequences of de-individuation in a group. *Journal of Abnormal and Social Psychology,* 1952, *47,* 383–389.

Freedman, J. L. *Crowding and behavior.* New York: Viking Press, 1975.

Freedman, J. L., Birsky, J., & Cavoukian, A. Determinants of behavioral contagion: Density and number. *Basic and Applied Social Psychology,* 1980, *1,* 155–161.

Freedman, J. L., & Perlick, D. Crowding, contagion, and laughter. *Journal of Experimental Social Psychology,* 1979, *15,* 295–303.

Freud, S. *Group Psychology and the Analysis of the Ego.* London: Hogarth Press, 1923.

LeBon, G. *The Crowd.* London: F. Unwin, 1895.

Redl, F. The phenomenon of contagion and "shock effect" in group therapy. In K. R. Eissler (Ed.), *Searchlights on Delinquency.* New York: International Universities Press, 1949.

Singer, J. E., Brush, C., & Lublin, S. C. Some aspects of deindividuation: Identification and conformity. *Journal of Experimental Social Psychology,* 1965, *1,* 356–378.

Turner, R. H. Collective behavior. In R. E. L. Faris (Ed.), *Handbook of Modern Sociology.* Chicago: Rand McNally, 1964.

Watson, R. I., Jr. Investigation into deindividuation using a cross-cultural survey technique. *Journal of Personality and Social Psychology,* 1973, *25,* 342–345.

Wheeler, L. Toward a theory of behavioral contagion. *Psychological Review,* 1966, *73,* 179–192.

Zimbardo, P. The human choice: Individuation, reason and order vs. deindividuation, impulse, and chaos. In *Nebraska Symposium on Motivation* (Vol. 17). Lincoln, Nebraska: University of Nebraska Press, 1969.

11 Illness as an Emergent Norm or Doing What Comes Naturally

Sidney M. Stahl
Purdue University

ABSTRACT

The sudden occurrence of outbreaks of sociogenic illness has only recently begun to receive scientific attention. Few reports have been published in the literature despite the apparent relatively high prevalence of such events. This chapter is an attempt to use social psychological factors and theory to understand such occurrences and to provide a set of theoretical explanations which can be generalized to other such events. To this end, labeling theory, emergent norms and coping theories are used. Examples from a variety of such events are used to integrate the explanation.

The intent of this study is to describe the occurrence of an outbreak of illness among a group of workers in a data processing center of a university and to attempt to explain this outbreak in social psychological terms. The use of social factors for an explanation is critical because no apparent biophysical or chemical agent could be found for the occurrence of the event. In addition, it is hoped that the beginning of the development of a theoretical model for the explanation of such phenomenon, which have a prevalence apparently greater than reported in the scientific literature will obtain. To present this explanation, several such occurrences already reported in the literature will be used to weave a theoretical fabric that can be applied in such additional outbreaks. The specific event in the Data Center has been published elsewhere (Stahl & Lebedun, 1974) and will be briefly reviewed below.

At the outset, a caveat on this attempt at explanation is necessary. The orientation of the discipline of medical sociology is one of explanation or understanding rather than amelioration. Hence, although it is hoped that this work is "relevant" to the problem at hand, medical sociology does not normally do research in an industrial setting and therefore does not have to concern itself with those problems that industrial psychologists, environmental scientists, physicians and managers must adjust to. In this work, the orientation is more toward explaining the sociogenic nature of the events than with deriving a solution for helping the worker or management of the industry solve the problem.

MYSTERY GAS: DESCRIPTION

The event described here happened in the Data Processing Center of a major midwestern university. The town's only industry is education thus limiting other employment opportunities of the Data Center workers. The sort of work done in the three story building consists of key punching and verifying records of all university salaries, purchase vouchers, student records and grades, and other routine university business. Because of the nature of the Data Center's work, few outside persons entered the building and the work was highly routinized. All affected workers were housed on the building's first floor in a single large room with about 30 old and noisy keypunches and an equal number of verifying machines.

The organizational structure of the keypunch operation is important to the event. The Data Center supervisor is housed on the second floor with the computers and other managerial personnel and the person directly responsible for the keypunch operation and its 35 workers sits in a glass enclosed office above, and in the corner of, the large and open keypunch room. From this vantage, she can observe all activity in the room. Directly under her supervision are two assistant supervisors who do some keypunching and verifying, assign tasks, and are expected to report any problems among the women who work in the room to the supervisor. One assistant supervisor is responsible for the keypunching operation, the other for the verifying operation.

Also important is the fact that the workers are expected not to converse during working hours about nonwork-related matters except at two 15 minute coffee breaks during the day and at lunch time. The coffee breaks are preassigned and staggered thus precluding interaction between friends except for those accidental periods during which two or more friends are assigned the same break. Even trips to the toilet were a subject of some concern regarding the rigidity of supervision on the part of some of the workers; one worker expressed the feeling that the supervisor was "aware" of the frequency of toilet use.

Although there was no contact with the public by the workers, the building supervisor dictated a rather strict dress code for his female employees. This was mentioned frequently by the workers who felt that the poor pay did not warrant the purchase of "business clothes" and that they could do their job just as well in any reasonable form of attire.

The occurrence of the "Mystery Gas," as it was headlined in the community's newspaper, took place on a Wednesday and Thursday in the early spring. At about mid-morning on Wednesday a majority (exact prevalence figures are only estimates) of the 35 workers in the keypunch room became ill with symptoms that included headaches, smelling a "strange and burning odor," hyperventilating, nausea, vomiting, burning and tearing eyes and fainting. Ten of the workers were taken to the university hospital's emergency room either by ambulance, by other workers or by supervisory personnel. The Data Center was closed for the day and environmental scientists began the task of taking multiple air samples.

On Thursday, the workers reported back to work and within one hour of opening the symptoms recurred. The building was again closed, workers returned to the emergency room or went home and the team of environmental scientists again performed air sampling through the day and well into the night.

The building was reopened on Friday. Prior to the beginning of work the supervisory personnel, a physician from the emergency room, and the professor in charge of the environmental studies met with the workers to assure them that whatever the problem was, it had now passed. They did not specify precisely what had caused the series of events but suggested that an "atmospheric inversion" had caused smoke from the adjacent university power plant to accumulate and descend upon the Data Center. It is critical to note that they did not tell the workers that the physiological evidence from the blood and urine testing and the air sampling produced no evidence of the effects of the hypothesized "atmospheric inversion." The workers accepted this explanation and work resumed in the Data Center with only one worker experiencing symptom recurrence later in the day. The one worker who began to hyperventilate was quickly isolated and sent home before others could react to her symptoms.

It was on Friday that it was possible to interview the workers who returned to work. Unfortunately, about one third of the workers absented themselves on this day; however, from reports of other workers, the symptomatology of those absent was no more or less severe than those who did return to work.

Precipitating Events

The work of Colligan and Murphy (1979), that of Kerckhoff and Back (1968), Knight, Friedman, and Sulianti (1965), and Levine (1977) all indicate that outbreaks of such events tend to follow some stressor event. This was clearly the case in the observed event. The university was in the process of building an addition to the adjacent power plant. Thus, heavy machinery was constantly being used and there were unanticipated (at least to the workers) dynamite blasts of bed rock throughout the day for a 2-week period prior to the event's occurrence. During the time in which the interviewing was being conducted in the building, at least four such blasts were heard and felt; the experience was at best disconcerting. One worker expressed the idea that she should be receiving a salary bonus for "hazardous duty." Furthermore, workers were being asked to work overtime, which many expressed as burdensome as most were married, several had small children, and few liked the job.

The most immediate event that seemed to trigger the outbreak was the moving of a large diesel engine under one of the open windows of the Data Center to run several pneumatic hammers. Although those women initially affected by the "mystery gas" were distributed throughout the room and not necessarily near the open window, the majority of the workers mentioned the presence of the engine. The environmental studies, however, were unable to detect noxious fumes from this source.

Methods

During this Friday the entire day was spent in the Data Center interviewing privately the workers and supervisors. The interviewing was both open-ended and by means of some quickly constructed interview schedules that had face validity. Several hypotheses were tested including the relationship between work satisfaction and symptom severity (an inverse relationship was expected), the empathy between the supervisor and the worker and the worker's level of job satisfaction (a positive relationship was expected, and found), and an exploratory hypothesis regarding the manner in which the symptomatology spread throughout the building. Interviewing lasted from the beginning of the work day until about four in the afternoon when the building supervisor confronted the author and his co-interviewer (Dr. Morty Lebedun), with the statement that he thought what we were doing was not advantageous to the morale of the workers (although workers were almost too anxious to talk with us), that our interviewing (which lasted for about 15 to 30 min per worker) was disrupting his ability to complete the day's work, and that we had probably collected "enough information."

Findings

It was found that those workers who were most dissatisfied with their jobs were the ones most likely to report the most severe symptomatology. Those who fainted and vomited were invariably the persons who had attempted to look for other jobs or expressed the highest levels of dissatisfaction with the work conditions, the nature of the relationship between the supervisor and the workers, and the restrictive dress code. These were also the persons who suspected that the supervisors kept track of the number of times the toilet was used, and stated that the supervisors were too punitive regarding the use of the pay telephone in the building. It is interesting to note that one of these workers had mentioned to her immediate supervisor that she was looking for another job and was verbally chastised by the first floor supervisor.

In addition, there was a strong association between the supervisor's perception of the worker's personality and the worker's level of job satisfaction. The supervisor was adequately "reading" the worker, at least in this regard, and the worker knew it. Fascinating was the fact that at no time was competence as a keypunch operator ever mentioned by the supervisor. All judgements seemed to be based on the personalities of the women employed in the Center.

Perhaps the most interesting finding is that persons who considered each other friends in the Data Center, shared symptoms. The construction of a sociogram, that is asking each worker to select other workers who they considered as friends, or to select those other workers whom they respected or talked with most frequently, indicated that the four women in one of the self-described sociometric networks all fainted and that no other worker in the Center fainted. Similarly, the five workers in another sociometric network were the only ones who vomited; no other workers shared this physical expression of the "mystery gas." In the same manner, two women, one a foreign person who spoke little English but had prior keypunching experience, said they became agitated when they saw the events occurring around them but experienced no symptomatology. They mentioned no friends in the Data Center and no one mentioned them in their sociograms.

Further elaborating this point, it is interesting to note that the workers were highly aware of the temporal order of events. Statements such as "first Sally started breathing funny and then Mildred's eyes started watering," or "Jane vomited and then I noticed that Ann also vomited and then I had to vomit," were made. The symptoms were essentially "managed" in such a manner that many workers were able to verify who got what, when and where, suggesting that the idea of "social contagion" (Milgram & Toch, 1969) was clearly operative in this situation.

Explanation

Several of the above findings require elaboration and explanation. The intent here is to attempt to explain in theoretical terms those events that were most significant and which should have some utility for understanding future outbreaks of such "mass" epidemics.

1. Gender: One of the most persistent variables in the "mystery gas" study reported here, as well as in other studies, is that only the women in the Data Center were affected by the event. Explanations for this, at least in the incident reported here, might be that only women were on the first floor of the building and/or that the men working in the building had both more varied and potentially interesting jobs. Thus, in the case under study, there is not an adequate control group.

However, in virtually every study of similar occurrences, the vast proportion of persons affected are female. Levine (1977) reports that 57% of those fainting in a high school band were girls; Knight et al. (1965) reports that all but one student who fainted in a southern school were girls; Kerckhoff and Back (1968) report almost all of the affecteds were women in the southern textile plant that they studied. In fact, in Colligan and Murphy's (1979) review of 16 separate incidents that they found in the literature, or for which they had access to the data, 93% of the affecteds were female. There are several potential explanations for the preponderance of females found in the affected category. Although somewhat interesting, the two most frequently found in the literature seem to beg the question.

The first explanation indicates that men are more goal oriented than women (Colligan & Murphy, 1979) and despite psychological strain, men are more likely to view the work task as a means to an economic end and therefore will ignore the strain and will not express that strain in terms of a stressor reaction, such as illness. In this rather sexist explanation, women are said to be more "emotional" or to better express their feeling states, whereas men are better able to control their emotional releases. The explanation is not only sexist, but is unsupported by empirical evidence. The goal orientation supposedly found in men more frequently does not explain why women can run sewing machines as well as men (Kerckhoff & Back, 1968) or play a wind instrument in a band (Levine, 1977) as well as their male peers. Goal orientation, by definition, is related to the task at hand, not some nebulous, ill-defined component of the environment.

A second explanation often posited is that for women, the work role is secondary to the wife/mother role. This supposedly provides a rationale for women seeking an easy role release from for-profit work roles. A reading of the analysis of work roles by Studs Terkel in his book *Working* (1972) and the work of Smelser (1963) and Bendix (1956), indicates that working is not terribly exciting for the average person, male or female, on the assembly line. Work is a place to be for 8 hours a day and both males and females seem to dislike repetitive tasks equally. This is further supported by the fact that hourly employees miss more work due to illness than do highly salaried professionals and that lower-income adults use health care services more frequently than higher-income adults (Monteiro, 1973). The fact that some females have another socially functional role to fulfill outside of the for-profit work force does not mean that they are more likely to report illness than males to seek secondary advantages.

Perhaps more to the point regarding the female preponderance in the prevalence of affecteds is the fact that women have traditionally known how to use the health care delivery system in more appropriate ways than men. The latest data available from the United States National Center for Health Statistics (1979) indicate that 47% of the women and 38% of the men in the United States had a "general checkup" within the last year; 80% of the females and 70% of the males saw a physician within the last year. Although use patterns are higher for women than for men, death rates in the two groups, even controlling for age, and especially among those still in the work force, are now and consistently have been lower for women.

Thus, women have been more adequately socialized in the use of the health care delivery system and it is more socially appropriate for women to use the system whereas for men, use of the system for complaints other than those perceived to have low commonality or thought to be serious, is not a socially sanctioned behavior. It is not surprising then, that in light of a stressor event, we find women more frequently somatocizing the stress than we do among men. Phillips and Segal (1969) make this point and conclude that women are more willing to discuss illness than are men. This of course leads to the greater prevalence of women in data on the number of physician visits per year and suggests the appropriateness of women using illness as a socially legitimate role release. Because the avenue of illness is appropriate, then it is more likely that they will use that avenue to provide perfectly legitimate releases from various social roles, such as the work role, which they find, for whatever reason, to be situationally problematic.

Phillips (1964) makes the same point in a different manner. He states that males are more stigmatized, or negatively sanctioned, than are women for using various forms of behavior, such as illness, as a role release. Thus, again, the social system sanctions the use of illness as a role release for women, whereas for men, a similar role release is more negatively viewed.

2. *Socioeconomic Status:* Like the variable gender, it is no random accident that most of the studies reported in the literature involve persons of lower and lower-middle socioeconomic status. Kadushin (1964) reports that illness prevalence in general is probably about equal in all social classes, with some obvious exceptions, but that the lower status individual "feels" more sick than does his/her more middle class counterpart. He postulates that in the lower classes anxiety is more frequently expressed in physical terms than in emotional terms. This may in fact be because the shared vocabulary of psychological distress is found predominantly in the middle class; a good deal of psychiatry, much to its collective chagrin, has become a middle class service phenomenon.

Crandell and Dohrenwend (1967) share this perspective in their excellent study by relating that lower class individuals express psychological distress in somatic terms whereas middle class individuals express the same types of psychological distress in psychological terms. This conclusion is also shared by McBroom (1970). Thus, it is possible that in the Data Center, those individuals who were more highly educated, one indicator of socioeconomic position, and who had the more prestigious jobs, such as the men responsible for computer programming and the coordination of all Data Center work, may have experienced the same stressors but were able to verbalize them. The lower status workers, who were all women, reacted in a pattern of somatocizing their distress.

3. Illness as Response to Strain: In the "mystery gas" experience described earlier as well as in virtually all events reported by Colligan and Murphy (1979), some form of precipitating event triggered the outbreak of the illness. The event in the Data Center consisted of the moving of a diesel engine under one of the windows following a long period of intense noise, dynamite blasts, the strain of imposed overtime and a very unpopular management style. One has the impression from this situation as well as from many of the other situations studied, that those persons in the affected category felt trapped by an intolerable structure and that something had to give. Lest one get the impression that this is a Freudian interpretation, some explanation is necessary.

There is in the area of social epidemiology, a long tradition of research and theory that attempts to relate social factors to the onset of physiological symptomatology and disease. Some of these studies include the following: changes observed in serum uric acid and cholesterol levels in men undergoing job loss (Kasl, Cobb, & Brooks, 1968); coronary heart disease attributable to occupational stress (House, 1974); normal life events seen as stressors leading to a catalogue of morbidities (Dohrenwend, 1973); spouseless motherhood and its relationship to elevated levels of physical and psychological incapacity (Berkman, 1969); the relationship between various psychosocial factors in the prevalence of such disparate signs as hypertension and lower back pain (Backus & Dudley, 1974); and social change leading to self-perceived disease in a Dutch population (Aakster, 1974).

These empirical observations exist within a broader theoretical underpinning although the various theoretical structures used to explain them are as varied as are the findings themselves. Most critical among the explanations are those by Hans Selye (1956), Moss's statement on *Illness, Immunity and Social Interaction* (1973) and the work of Dohrenwend & Dohrenwend (1974). Generally, these statements postulate that the extent to which the social environment is seen as stressful will lead to the occurrence of both illness and disease events. Thus, a positive relationship can be expected between the amount of strain in the social environment and the production of a stress, which in Selye's terms is expressed physiologically.

If, as in the "mystery gas" experience, other avenues of amelioration are blocked, illness may become an appropriate stress reaction to the environmental strain. Furthermore, and independently of the blockage of legitimate avenues of change, the studies relating social factors precipitating illness and disease indicate that a physiological reaction is a normal outcome of the presence of a stressor; that is Selye's major thesis (1956). Hence, fainting, vomiting, hyperventilating, and the group of nonspecific symptoms reported in the "mystery gas" experience as well as in other reported events, are perfectly appropriate and socially legitimate ways of excusing oneself from a bad situation.

Mechanic and Volkart (1960) further elaborate this point by finding that illnesses and symptoms that are common, relatively familiar to the group, and have predictable and nonlife threatening outcomes are much more likely to be used in an attempt to gain access to the sick role. Certainly the list of symptoms reported in the various studies reviewed by Colligan and Murphy (1979) fall in this common, nondangerous and predictable category. Hence, the sick role (Parsons & Fox, 1952) through use of these symptoms, is an appropriate role release. Illness is, essentially, a coping mechanism.

4. Selective coping: If we accept the proposition that illness and/or symptomatology is a coping mechanism, then it is appropriate to ask why it is that not all persons react to an obviously negative set of stressors. Some persons in the "mystery gas" experience developed no clinical signs and in fact did not even smell the "mystery gas" whereas others were taken unconscious to the hospital's emergency room. The same variability of reaction is found in virtually all of the events that Colligan and Murphy (1979) reviewed.

There are probably two basic reasons for this less than perfect relationship. One will be elaborated later in a discussion of the meaning of social support systems. The other explanation of the nonperfect nature of the relationship between stressor and reaction is the fact that, while on the surface, the stressor events are less than ideal situations within which to work, not all individuals perceive them as being equally problematic. The classic epidemiological model postulates that there is a linear relationship between the stressor event usually found in the social structure and the disease outcome (Stahl, Grim, Donald, & Neikirk, 1975). The search for a direct relationship between agent and outcome used so frequently in social epidemiology comes from the monoetiological model of disease. Disease is said to be explained by looking only at those persons exposed to a given infectious agent. Although this may constitute a necessary relationship, it ignores the sufficiency explanation in the philosophy of science (Nagel, 1961). In some recent work on hypertension (Stahl et al., 1975) it is postulated that those individuals who perceive the structural events as inhibiting life goals are the persons who will eventually contract the disease. Translated in terms of the "mystery gas" experience and applied to a number of studies reviewed by Colligan and Murphy (1979), it appears that only those persons who perceived that the stressors were severe, and/or socially disruptive, were the ones to react by developing some form of symptomatology. This explanation is isomorphic with the data in that those persons who were most dissatisfied with the work situation at the Data Center were also the ones to develop the most severe symptomatology. Those who used illness as a legitimate role release were the ones who perceived the structural situation to be so severe that they had looked for another job within the last year, felt that the supervisors were grossly unfair, and generally disliked the work situation. These were also the individuals who felt that they simply had a "job" and that their work prevented them from spending time at other important social obligations, such as their families.

Richard Lazarus (1978) expands on this thesis by applying it directly to hypertension. He postulates that coping is based on the specific environmental and psychological events and that without an understanding of the individual's ability and willingness to cope with these events, it is impossible to understand the selective reaction of the various participants following the stressor event.

5. Social Support System: Perhaps the most fascinating aspect of the "mystery gas" experience is the manner in which the illness was "transmitted" throughout the group. Transmission relates directly to the selective coping process just elaborated. Mentioned above was the fact that not all persons experienced the illness and that this might be due to the selective perception of the severity of the potentially stressor events. The lack of a perfect relationship between the stressor events and the onset of the illness may also be due to the fact that persons are less than perfectly integrated into the social support system or are isolates in the information network within the work site (Moss, 1973). In the "mystery gas" experience, it was noted that those persons who were isolated from the friendship and communication patterns within the Data Center experienced no symptoms. On the other hand, those persons involved in specific informal information and friendship patterns shared a common set of symptoms. The intent here is to explain the nature and meaning of these friendship patterns for the occurrence of the outbreak of the sociogenic illness.

Social support systems are those networks of friends and/or primary relationships that provide role models for one's behavior and which lend social credibility and positive sanctions to the behavior of the individual. Thus, a social support system might be a family, one's colleague group, or the work group within which one operates for 8 hours a day. Kaplan, Cassel, and Gore, (1977) provide a useful framework for the study of the effects of social support networks. Although research in this area is relatively new, it does seem to indicate that the existence of social support networks is significant in moderating the effects of stressful environmental stimulae as well as providing the appropriate types of cues for responding to those stressors (Gore, 1978; Antonovsky, 1979).

It is this latter point that will be argued here: The availability of role models for illness provides the basic mechanism through which sociogenic illness can "spread." That is, because of the availability of social support networks among peers, the Data Center workers essentially "learned" how to behave in the stressful situation.

Turner and Killian (1972) develop the concept of "emergent norms" which can be used to explain the phenomenon of the spread or contagion of the series of symptoms observed in the "mystery gas" experience as well as in a number of the other outbreaks reviewed by Colligan and Murphy (1979). That is, in any collective behavior, whether it is an antiwar demonstration or a number of workers sharing a common location, milling behavior must precede any collective action. Milling behavior is such that persons circulate among one another and share their perceptions of the events that are occurring. From these shared perceptions, definitions of the situation arise which are communicated to others in the environment. These shared communicated perceptions then emerge as the basis for norms regarding how to cope or handle the situation at hand. In this way, the affected group shares a common definition of the situation and establishes a series of normative behavior patterns that are unique to the given situation. Because the situation under investigation here is itself rather abnormal, the emergent norms are not necessarily congruent with the norms that exist for the larger social system. Hence, the norms that emerged in the "mystery gas" incident might have been negatively sanctioned by the supervisors but were considered appropriate and useful by those individuals trapped in the situation. A shared, collective response was established regarding how to react to a situation, which consisted of the accumulation of a long-standing grievance addressed to the social structure and exacerbated by the intolerable (as perceived by the workers) work conditions.

Stating that a norm for behavior emerged from the situation partially begs the question of why specific illness symptomatology was chosen by the participants to express the strain rather than coping by more legitimate means (as defined by the larger social system). Colligan and Murphy's (1979) review of the various studies of "mass psychogenic illness" indicates a remarkable similarity of symptomatology. These include: headache, dizziness, weakness, nausea, burning eyes, shortness of breath, vomiting, fainting, headache, hyperventilation and skin rash. Every one of these symptoms, or more appropriately clinical signs, has either been experienced by virtually every human or if not experienced, virtually everyone knows how to imitate or recognize the specific sign. In short, there is a high commonality of symptomatology of this nature. Nothing is terribly unique in any of the situations.

To explain this commonality, it is necessary to use part of a theory called labeling theory. Because this theoretical construct has generally been applied to criminal behavior and mental illness, it is critical to make perfectly clear that there is no intent in its use to draw any analogy between those deviant social behaviors and the occurrence of a sociogenic outbreak of illness.

In Scheff's classic work on mental illness (1966), a series of propositions supported by empirical evidence are presented. First, the various signs of mental illness, and for our purposes, the various signs repeatedly found in the outbreaks of sociogenic illness, are clearly quite common and easily learned. There is a stereotypic swooning pattern; everyone knows what being nauseated feels like and the mere suggestion of nausea triggers that same perception in others (remember dissecting your first laboratory frog in high school?). Thus, everyone learns these clinical signs. The nature of the immediate situation, being anxiety laden and very confusing to the participants forces them to look for an explanation of the immediate events. Because of the need for an explanation, the reaction of the immediate peer group becomes critical in determining both the duration and the outcome of this initially ambiguous set of stimulae. The reaction can turn to an elaboration of these easily learned and frequently previously experienced sets of symptoms.

Another proposition suggested by Scheff is that as in mental illness, there may be a reward for playing out these stereotyped behaviors. The victim now shares the experience with others thus making the behavior a collective one and reinforcing the actor's own perception of the reality of the event. The victim now has an explanation for her own feelings of anxiety that is shared by other participants. The attention given the affected workers, although absolutely justified in such situations, further reinforces the behaviors since others in the situation are acting as though the situation were "real." It is critical here to emphasize that this statement is meant to be "value free"; a lack of reaction by others in the situation, such as the supervisors who took workers to the hospital, would be both unethical and immoral.

In this way group participants develop a collective definition of the situation where few if any clues to what is happening are present. General dissatisfaction, triggered by some event such as the diesel engine and dynamite blasts, produces a vague and amorphous feeling state. This feeling state requires some definition as all people understand their social world by placing events and feelings into categories. Thus, a collectively shared and understood norm emerges to define the situation and persons proceed to do what comes more or less naturally. Illness is a perfect reaction and a quite legitimate manner of defining the situation and reacting to it. Illness provides a set of parameters, easily shared by others, which can provide immediate relief from the collectively experienced situational difficulty. Furthermore, McEvedy and Beard (1973) report that affecteds report more sick days from work and a greater prevalence of hospital admissions than do nonaffected workers. This implies, then, that affecteds already know how to use the system and how to "experience" illness in a manner differently than do nonaffected workers.

These explanations are further strengthened by looking again at the sociometric analysis that emerged from the "mystery gas" experience. Using the sociogram technique, which asks the workers to specify other workers in the Data Center considered to be friends or persons with whom they spent most of their time at work, a pattern emerged that further explained the spread of the sociogenic illness. All persons within a given sociometric network shared almost identical symptoms and few workers in other networks had those symptoms. Thus, there was a network of fainters, a network of hyperventilators, etc. Where one individual had multiple symptomatology, the most severe of the symptoms (usually fainting) was used as the indicator.

It is hypothesized that even within these small sociometric networks, the victims "learned" the appropriate response to the noxious situation. The norm for behavior emerged by individuals taking cues from others with whom they normally shared interaction. In this way, a study of the social relationships showed a high prevalence of "appropriate" symptomatology. Furthermore, workers within each network generally knew the temporal pattern of events which preceded their own contracting of the appropriate symptomatology. Generally, the person most frequently selected as the center of the sociometric network (called the "sociometric star") preceded other members of the network in developing the "appropriate" symptomatology which was subsequently "shared" with other sociometric members.

This pattern more closely follows that of contagion theory rather than convergence theory (Milgram & Toch, 1969). In convergence theory, affecteds seem to develop symptoms or responses more or less simultaneously. A detailed questioning of the affecteds in the "mystery gas" experience indicates that the workers were quite aware of the pattern of events. Those persons who developed a response to the situation communicated these responses, during the milling phase of the collective behavior and thus a norm for the specific illness sign emerged within each sociometric group. The question remains, of course, as to why one symptom was used in one sociometric group and why another emerged in a different group. These innovations were adapted by the leaders in the situation and seemed to spread to others in the group (Coleman, Katz, & Menzel, 1966).

CONCLUSIONS

The sociogenic outbreak of illness was stopped at the Data Center when sources external to the workers and with socially legitimate expertise provided an explanation for the series of events. The environmental scientists and physicians essentially ended the outbreak by declaring that the "atmospheric inversion" was responsible, despite a lack of clear evidence that this phenomenon either occurred, or if it did, had any detectable physiological effect on the workers. Thus, the "mystery gas" was given a label and the behavior of the workers was legitimated.

It is rewarding to know that the supervisors of the Data Center subsequently reevaluated their management policies and instituted changes in the organization of both the work and the severity of supervision. As far as this author is aware, no additional problems have occurred.

Several points must be mentioned in concluding. One of these points bears on the ethical nature of such research. Without a sympathetic audience among the management of the Data Center, the results for the workers might have been disastrous. That is, inasmuch as the workers were not unionized, and keypunch operators in a university population are abundant, there is the possibility that these people might have been fired for being disruptive to the ongoing operation of the Data Center. In making reports exclusively to management, which Colligan and Smith (1978) state frequently happens in such situations, the investigators should remain highly sensitive to the implications of their findings. To "blame" the workers for hysteria or malingering because no toxic agent can be found is both unwarranted and unethical.

Hopefully, the reader has noticed that the term "psychogenic" illness has been avoided and instead "sociogenic" has been used in describing the occurrence of events. This is meant to be more than a bit of chauvinism from a sociologist. The events explained here, as well as those analyzed in other social settings, have as much group or peer input as they have psychological explanations. In fact, one of the central theses presented here is that the norms of behavior emerge from the group and that to explain the event requires an understanding of both psychological characteristics of the participants as well as an interpretation of the group dynamics, which is after all, the immediate environment for the fostering of the event.

Finally, and related to the above issue, is one of the major drawbacks in all of the data reported in the literature. Looking for sociogenic or more broadly put, social psychological pathogens, in the occurrence of these events has consistently been relegated to a residual category of research. That is, if standard biochemical techniques fail to explain the event, it is only than that the social scientists are called upon. This probably results from two factors: those "in residence" who study such phenomenon are in fact, biologically or physically trained scientists and secondly, outbreaks of these events call for the immediate amelioration of the problem, not the more complicated, long-term understanding and explanation of the situation which can be more readily provided by the social scientists. Colligan and his colleagues (Smith, Colligan & Hurrell, 1978) are highly aware of this problem and have developed a series of social psychological instruments which can be used cojointly with biophysical techniques usually used in these situations. This development is critical for a better understanding of the occurrence of such events.

The convergence of the social and biophysical sciences in this area, as in other areas such as clinical medicine, will represent the benchmark for a scientific future which can better approach an understanding of humankind. The models presented here and the ongoing research in this area should better explain why it is that people, faced with a given situation do what comes naturally.

REFERENCES

Aakster, C. W. Psycho-social stress and health disturbances. *Social Science and Medicine,* 1974, *8,* 77–90.

Antonovsky, A. *Health, stress and coping*, San Francisco: Jossey-Bass Publishers, 1979.

Backus, F. I., & Dudley, D. L. Observations of psychosocial factors and their relationship to organic disease. *International Journal of Psychiatry in Medicine,* 1974, *5,* 499–515.

Bendix, R. *Work and authority in industry. Ideologies of management in the course of industrialization.* New York: Harper & Row, 1956.

Berkman, P. L. Spouseless motherhood, psychological stress, and physical morbidity. *Journal of Health and Social Behavior,* 1969, *10,* 323–334.

Coleman, J. S., Katz, E., & Menzel, H. *Medical innovation: A diffusion study.* Indianapolis: Bobbs-Merrill, 1966.

Colligan, M. J., & Murphy, L. R. Mass psychogenic illness in organization: an overview. *Journal of Occupational Psychology,* 1979, *52,* 77–90.

Colligan, M. J., & Smith, M. J. A methodological approach for evaluating outbreaks of mass psychogenic illness in industry. *Journal of Occupational Medicine,* 1978, *20,* 401–402.

Crandell, D. L., & Dohrenwend, B. P. Some relations among psychiatric symptoms, organic illness, and social class. *American Journal of Psychiatry,* 1969, *123,* 1527–1537.

Dohrenwend, B. S. Life events as stressors: A methodological inquiry. *Journal of Health and Social Behavior,* 1973, *14,* 167–175.

Dohrenwend, B. S., & Dohrenwend, B. P., Eds. *Stressful life events: Their nature and effects.* New York: Wiley, 1974.

Gore, S. The effect of social support in moderating the health consequences of unemployment. *Journal of Health and Social Behavior,* 1978, *19,* 157–165.

House, J. S. Occupational stress and coronary heart disease: a review and theoretical integration. *Journal of Health and Social Behavior,* 1974, *15,* 12–27.

Kadushin, C. Social class and the experience of ill health. *Sociological Inquiry,* 1964, *34,* 67–80.

Kaplan, B. H., Cassel, J. C., & Gore, S. Social support and health. *Medical Care,* 1977, *15,* 47–58.

Kasl, S. V., Cobb, S., & Brooks, G. W. Changes in serum uric acid and cholesterol levels in men undergoing job loss. *Journal of the American Medical Association,* 1968, *206,* 1500–1507.

Kerckhoff, A. C., & Back, K. W. The June bug: A study of hysterical contagion. Appleton-Century-Crofts, New York, 1968.

Knight, J. A., Friedman, T. I., & Sulianti, J. Epidemic hysteria: A field study. *American Journal Public Health,* 1965, *55*: 858–865.

Lazarus, R. S. A strategy for research on psychological and social factors in hypertension. *Journal of Human Stress,* 1978, *4,* 35–40.

Levine, R. J. Epidemic faintness and syncope in a school marching band. *Journal American Medical Association, 1977, 238,* 2373–2376.

McBroom, W. H. Illness, illness behavior and socioeconomic status. *Journal of Health and Social Behavior,* 1970, *11,* 319–326.

McEvedy, C. P., & Beard, A. W. A controlled follow-up of cases involved in an epidemic of benign myalgic encephalomyelitis. *British Journal Psychiatry,* 1973, *122,* 141–150.

Mechanic, D., & Volkart, E. H. Illness behavior and medical diagnosis. *Journal of Health and Human Behavior,* 1960, *1,* 86–94.

Milgram, S., & Toch, H. Collective behavior: Crowds and social movements. In G. Lindsey & E. Aronson (Eds.), *Handbook of Social Psychology, Vol. 4,* 2nd ed. Reading, Mass: Addison-Wesley, 1969.

Monteiro, L. A. Expense is no object . . . : Income and physician visits reconsidered. *Journal Health and Social Behavior,* 1973, *14,* 99–115.

Moss, G. E. *Illness, immunity and social interaction. The dynamics of biosocial resonation.* New York: Wiley, 1973.

Nagel, E. *The structure of science problems in the logic of scientific explanation.* New York: Harcourt, Brace and World, 1961.

Parsons, T., & Fox, R. Illness, therapy and the modern American family. *Journal of Social Issues,* 1952, *8,* 31–44.

Phillips, D. L. Rejection of the mentally ill: the influence of behavior and sex. *American Sociological Review,* 1964, *29,* 679–687.

Phillips, D. L., & Segal, B. E. Sexual status and psychiatric symptoms. *American Sociological Review,* 1969, *34,* 58–72.

Scheff, T. J. *Being mentally ill: A sociological theory.* Chicago: Aldine, 1966.

Selye, H. *The stress of life.* New York: McGraw-Hill, 1956.

Smelser, N. J. *The sociology of economic life.* Englewood Cliffs, N.J.: Prentice-Hall, 1963.

Smith, M. J., Colligan, M. J., & Hurrell, Jr., J. J. Three incidents of industrial mass psychogenic illness. *Journal of Occupational Medicine,* 1978, *20,* 399–400.

Stahl, S. M., & Lebedun, M. Mystery gas: An analysis of mass hysteria. *Journal of Health and Social Behavior,* 1974, *15,* 44–50.

Stahl, S. M., Grim, C. E., Donald, C., & Neikirk, H. J. A model for the social sciences and medicine: The case for hypertension. *Social Science and Medicine,* 1975, *9,* 31–38.

Terkel, S. *Working: People talk about what they do all day and how they feel about what they do.* New York: Pantheon Books, 1972.

Turner, R. H., & Killian, L. M. *Collective behavior.* 2nd Ed. Englewood Cliffs, N.J.: Prentice-Hall, 1972.

United States National Center for Health Statistics. *Physician visits volume and interval since last Visit United States–1975.* Series 10-Number 128. DHEW Publication No. (PHS) 79-1556. Office of Health Research, Statistics, and Technology, Public Health Service, U.S. Department of Health, Education and Welfare, Hyattsville, Maryland, 1979.

12 A Social Psychological View of Mass Psychogenic Illness

Alan C. Kerckhoff
Duke University

ABSTRACT

The perspective described here is most applicable to epidemics in closed populations, such as schools or factories. It postulates a situation in which there is a high level of collective tension, the introduction of one or more cases of dramatic individual illness events, and the resulting increase of tension and associated physiological symptoms. Those who become ill can be differentiated from those who do not according to level of tension, personality characteristics and position (spatial and social) in the situation. The importance of a credible explanation of the illness is emphasized, and the actions of experts who seek to understand the problem are seen as sources of validation of the credibility of the danger. These actions are also seen as contributing to a tendency for later cases to be less severely ill and to the ultimate dampening of the contagious response. Reasons for the preponderance of female victims are discussed, and several kinds of issues future investigations might emphasize are suggested.

INTRODUCTION

Although reports of mass psychogenic illness have been available for centuries, only during the past few decades has there been any significant scientific interest in them. The increased scientific interest has resulted in part from the recognition

Note: The contents of this report are reproduced herein as received from the contractor. The opinions, findings, and conclusions expressed herein are not necessarily those of the National Institute for Occupational Safety and Health, nor does mention of company names or products constitute endorsement by the National Institute for Occupational Safety and Health.

that such cases are frequent in occurrence and "expensive," both socially and economically. Perhaps even more important in explaining the increased interest, however, is the greater acceptance of psychogenic explanations of illness behavior. This has led investigators to go beyond identifying such epidemics by the exclusion of other sources of illness and to attempt to specify the dynamics of the process by which they develop. This chapter discusses one view of such epidemics of mass psychogenic illness.

The critical feature of the social psychological perspective used in this analysis is its focus on the interpersonal and cultural context within which mass psychogenic illness develops. It recognizes that those examples which have received the most attention recently all occur in relatively constrained social settings in which the participants are limited in the degree to which they can determine on their own the kinds of behaviors they will engage in. This structured context is especially important in understanding what generates such psychogenic epidemics. Equally important in understanding them is a set of cultural definitions which specify appropriate methods by which individuals may cope with their personal problems and their social relations with others in those settings. The definitions that prove to be important include both those which pre-exist in the group involved and those that evolve during the period of disruption.

The view discussed here may be summarized as follows: Mass psychogenic illness tends to occur in a situation in which a number of people find themselves exposed to sources of strain from which no fully acceptable means of escape is available. Those who are in the situation may know that some actions would serve this purpose, but the actions would also lead to negative outcomes, and they are thus not used. The combination of the original strain and the difficulty of using any known means of coping generates a general state of tension whose level is likely to be raised through interstimulation among those present. The state of unresolved tension leads at least some people to experience physiological symptoms. These symptoms come to be associated in the minds of at least some of those present with some credible, though factually incorrect, "cause" in the setting. Such an external "cause" objectifies the source of the symptoms and adds a further source of strain—the fear of the threatening agent. The belief in the threatening agent, the increasing number of victims, and the actions of outside experts which tend to validate the credibility of the threat all serve to increase the probability that others will experience similar symptoms which may be defined as being due to the presumed source of the spreading sickness. (See Kerckhoff, 1970 for a more general statement of this perspective.)

In the process of spelling out the details of this social psychological perspective, the chapter cites examples of the kinds of processes discussed. It then turns to the prospects for further work on the problem, and asks: What kinds of research appear to be most promising to increase our understanding of cases of this kind?

THE DYNAMICS OF EVOLUTION

The Setting

Although it is possible to analyze epidemics of mass psychogenic illness within a broader framework than will be used here (see Markush 1973, Sirois 1974), the present discussion will be limited to examples occurring in relatively closed populations. Most of the examples appearing in recent writings have occurred in institutional settings, such as in a school or a factory. It seems likely that the present discussion could be extended to apply to other less circumscribed settings, but to do so would require us to discuss situations about which little empirical evidence exists. As will become apparent in what follows, theoretical leaps are needed even to discuss the more circumscribed settings; much greater gaps exist in our knowledge of more open situations.

There are two immediately obvious characteristics of such settings as schools and factories:

1. They involve relatively large numbers of people, all of whom are in roughly the same position in the social structure. Both schools and factories have large numbers of relatively low status persons (students, workers) who are defined as a homogeneous population by their superiors (teachers, supervisors).
2. In order to solve the functional problem of continuous and orderly operations, a number of limiting rules are established which restrict the amount of independent action on the part of all such low status persons. Workers can only take scheduled "breaks," students need a "hall pass," etc. Even the most enlightened management or administration must restrict free movement to a considerable degree, and it is impossible to accommodate to the full range of individual differences. This is the normal state of affairs in a school or a factory.

Such restrictions are, of course, variable, both from place to place and from time to time in a given institutional setting. Many of the studies of mass psychogenic illness have suggested that the epidemics occur under conditions of additional restriction or stress. For instance, pressure from management to increase production, or the introduction of a new procedure, or the requirement that workers work overtime can make the normally tolerable conditions more stressful. It is also true that most institutions have mechanisms for dealing with the normal tensions associated with the necessity of restricted individual autonomy. In the school, the teacher attempts to be sensitive to the individual needs of students, and counsellors and others serve as a further source of amelioration. In a factory, supervisors will attempt to balance the need for production with a humane concern for the workers, and unions and other worker organizations provide channels through which difficulties can often be resolved. Yet, these mechanisms do not always function satisfactorily from the perspective of the worker or student. Such organizations can become disorganized, individual supervisors or teachers can fail to meet others' needs, or external conditions (e.g., the pressure of production schedules, the upcoming examination) can limit what can be done to ease the problems experienced by the workers or students.

Many of the epidemics studied seem to have occurred in settings in which the pressures on the low status persons were greater than usual (Colligan & Murphy, 1979). A general sense of tension was experienced by the workers. A further condition also seems necessary if some form of mass psychogenic illness is to develop out of such tense situations. This is the inability of the workers who experience the tension to take appropriate action to relieve it. In some examples, for instance, inadequate avenues of redress within the organization may become important (Kerckhoff & Back, 1968). However, one obvious method of reducing the stress is always available—one can simply leave the situation. Clearly, in those situations in which a mass psychogenic illness epidemic occurs, that method has not been used. Workers get sick on the job, students get sick in school. There are two possible explanations for the fact that those who experience stress in such situations do not leave. First, in many such situations, especially for workers, there are obvious rewards associated with staying, and a loss of those rewards is associated with leaving. Absenteeism leads to loss of pay, and it can even lead to dismissal. Many workers remain on the job out of fear of such loss, thereby increasing the sense of strain. Second, if at least part of the job-related stress is due to unsatisfactory channels of amelioration on the job, any attempt to seek approval of a lessening of stress may even worsen the workers' relations with supervisors. There is therefore likely to be an uneasy balance between the need to do something about the problem that is faced and the need to avoid the kinds of action that seem most suitable for coping with the problem.

The discussion thus far appears to describe a situation in which workers or students can analyze the pros and cons of taking action to resolve a problem but are caught in a no-win situation. This really oversimplifies the matter. Actually, the source(s) of tension may not be clearly understood by those who experience it. Especially in well-established situations, the requirements may seem so natural and appropriate that it is not clear to those who experience them just what the problem is. There is often a pervading sense of tension without an understanding of its source. Most workers recognize that limitations, rules, and specified procedures are needed if the factory is to function adequately. They may not like it, but they will probably see such restrictions as something they have to accept if they are to hold that job. That may not reduce the tension, but it may reduce their ability to conceive of a way of lessening it. It is also important to recognize that if mass psychogenic illness is to result, tension must be relatively widespread in the situation. This does not mean that everyone must be equally distressed, but it does mean that the shared conditions must result in a shared sense of tension.

ILLNESS AND ITS "CAUSE"

A situation in which many people working in the same setting share a sense of tension that they do not understand and/or know how to reduce is undoubtedly fairly common. What we find is that in some such settings, one or more cases of illness appear, and these are explained (either by the victims or by others) as being due to some toxic element in the situation. The element may be a gas, a chemical, a poisonous insect, etc. The presumed toxic element must have two important characteristics, however. First, it must be *in* the situation. The early cases of illness must be attributable to something to which other workers are also presumably exposed. Second, it must be a *credible* threat. That is, it must be consistent with the general cultural beliefs regarding how people get sick (e.g., a toxic chemical is more credible for most Americans than witchcraft). In some cases, there is clear evidence to support the idea of the threat. In others, the presumed threat is the kind for which such evidence is difficult to produce. In some epidemics, a strange smell is defined as a poisonous gas. In others, a tiny insect or an irritant in the materials being handled or an unknown contagious disease is blamed for the illnesses.

This last characteristic, the kind of evidential basis for the explanation, is the most problematic. The epidemics reported in the press and even in the professional literature are often unclear about how the affected population explains the outbreak. It seems likely that where the reports are silent on this matter it is because of limited information rather than a lack of an explanation in the affected population, but we cannot be sure about that. It is clear, however, in those cases where an explanation is reported, that it may either be available before the first cases of illness or be provided after they occur. For instance, in some cases, it is clear that someone identified a "strange smell" or other foreign element in the situation prior to the first cases, and those cases were immediately attributed to that element (Roth, 1965). In others, the cases are initially attributed to a "mysterious malady" and only later does a coherent explanation evolve (Levine, Sexton, Romm, Wood, & Kaiser, 1974). It is difficult to generalize from present evidence beyond indicating that once a number of cases of illness occur, they generally come to be linked together as having been due to the same threatening force, a force that is in the school or work setting and which continues to threaten others in that setting. In most cases, some kind of "cause" of the cases comes to be agreed upon by many people in or associated with the situation, but even where the cases remain mysterious, they are seen as all being caused by "it"—whatever "it" is.

CONFIRMING THE THREAT

The most common, and obviously necessary, action taken by officials when the first case or cases of illness occur is to remove the individuals from the setting to where they can receive medical attention. Though this is certainly appropriate, it has an effect that goes beyond the simple medical treatment provided for the stricken individual; it calls attention to the cases, and it provides confirmation that the individuals *really are* ill. In the wake of a series of cases, other official actions are usually taken which even more dramatically validate the significance of the threat. The plant or the school may be closed, experts are brought in to conduct various tests which seek to specify the source of the difficulty. If a gas is suspected, air samples are taken. If a toxic chemical is suspected, all kinds of tests may be conducted. If a poisonous insect is suspected, exterminators may be brought in. Such actions also tend to confirm the possibility of a serious threat.

This is where the credibility of the "cause" that has been suggested within the affected population becomes critical. If the kind of illness experienced by the early cases is a possible outcome of the kind of threat reported, officials will take the reports more seriously and will mobilize very visible and technically adequate means of seeking evidence of the threat. Even if the evidence is not found, the search itself increases the general credibility of the threat and makes it likely that some of those in the affected population who were at first skeptical will have second thoughts.

There is also a reciprocal reinforcement process involved in the validation of the threat. One or two cases of illness might not impress many of those present. However, as the number of cases increases, it is difficult to avoid believing that *something* must be making those people sick. Again, if a credible explanation involving a toxic element in the situation has been suggested, the more cases that occur, the more difficult it will be to discount that particular explanation. Also, the more an individual believes in the reality of the threat, the greater the tension (s)he is likely to experience, and the more likely (s)he is to experience symptoms associated with tension. Such symptoms are then very likely to be associated with the suggested threat, and the individual may well become a "case."

SUMMARY

The general view of mass psychogenic illness adopted here sees the individual cases as exhibiting symptoms that are associated with unresolved tension. It is assumed that many circumscribed settings, such as those found in schools and factories, impose sufficient restrictions on the free activities of individuals that many persons in those settings will share a sense of tension, and that some of those settings will be particularly tension-filled. The introduction into such a situation of one or more cases of illness, especially if the illness is dramatic (e.g., fainting), will suggest that a tangible source of threat is present. This would be especially likely in a population whose members do not recognize the possibility of a psychogenic contribution to discomfort and who, therefore, tend to "somatize their difficulties" (Bart, 1968). Those who are experiencing unresolved tension will be likely to associate their physical experiences with the presumed source of threat and attribute their discomfort to it. Belief in the presumed threat will increase the level of discomfort they experience and make it likely that other cases will occur and that they will be explained as being due to the same threatening element. The actions of officials, though necessary, will tend to validate the seriousness of the threat, thereby adding to the tension and increasing the likelihood that symptoms will be experienced and that they will be attributed to "it," whatever "it" is.

VARIATIONS IN TIME AND (SOCIAL) SPACE

If the general pattern of evolution of mass psychogenic illness just outlined is at least partially correct, it leads us to look for certain patterns in the development of such epidemics. It is also true that numerous studies of such epidemics have uncovered some patterns that can be used to raise further questions about the perspective just outlined. For instance, in no known example of mass psychogenic illness have all persons in the population involved become ill. This raises questions about channels of "contagion" and the possibility of "immunity." Also, if the explanations of the threat must be "credible," this raises questions about differences in the social definitions of what is credible and how they might affect the development of an epidemic. And, if the first (few) cases are presumed to be almost random events which act as initiators of a complex process in which a belief in a threatening force evolves, and if that belief provides an explanation of the symptoms others experience and leads them to become more seriously affected, this suggests that the early cases are different from the later ones. Finally, there has been a strikingly consistent pattern reported that suggests that women are much more likely than men to be victims of mass psychogenic illness. Some attempt must be made to assess the validity and significance of that sex difference.

WHO BECOMES A CASE?

Although the question that heads this section is central to understanding and attempting to deal with cases of mass psychogenic illness, no simple answer can be given at this time. However, the perspective proposed in this chapter suggests some directions in which future investigations might go in pursuing an answer. The general conceptualization presented earlier points to several factors that increase the likelihood of an epidemic and therefore increase the likelihood that any given individual will become ill. These factors are: tension experienced in the work (or school) situation, the experience of symptoms resulting from that tension, the belief that those symptoms and serious illness can result from a threatening element in the situation, and the failure to take adequate coping action to relieve the symptoms. Although all of these elements are included in the general theory of mass psychogenic illness proposed here, each of them is a variable rather than a constant, and thus various individuals in the situation will exhibit them to differing degrees. The probability of any given individual becoming a "case" in an epidemic of this kind should thus vary with the extent to which each of these factors characterizes him or her. There is evidence that variation on each of them is important in becoming ill.

The kind of tension relevant to mass psychogenic illness which has been emphasized earlier is the tension generated by the conditions experienced in the setting in which the epidemic occurs. Because the setting is not wholly uniform, undoubtedly people in different groups vary in their experiences. Yet, it is difficult to assess the significance of situational sources of tension without also considering experiences outside that setting. For instance, although required overtime work may be stressful for all workers, this stress varies according to the other kinds of demands on a worker's time imposed by outside groups—such as family obligations. For some, overtime simply means additional energy expended for a higher rate of pay. For others, overtime means very complicated arrangements for child care, club or church obligations, car pool schedules, and so on. Also, of course, there are variations in the kinds of stress experienced by workers outside the work place. For some, the tensions at work are simply added to many others experienced elsewhere. Without attempting to propose a "calculus of tension," therefore, it seems reasonable to assume that individuals in the same setting will actually be experiencing different levels of tension. An adequate understanding of the differential response to epidemics of mass psychogenic illness must certainly take that variation into account.

Many studies have attempted to assess the significance of personality characteristics in the distribution of cases within a population (see Colligan & Murphy, 1979). Basically, in the context of the present discussion, this suggests that individuals may vary in their ability to cope with tension or stress. Even holding level of tension constant, individual responses to it may vary. For instance, the tendency to attribute one's experiences to some external source rather than viewing them as being within one's own control, or the tendency to seek assistance rather than attempt to cope on one's own, seems likely to affect how an individual will react to a sense of tension in a work or school situation. In particular, how an individual responds to physical symptoms (fatigue, headaches, nausea, feeling faint, etc.), in fact the willingness to acknowledge that "something is wrong," may influence how (s)he will behave in such a situation. One very tangible way in which such differences can affect the development of an epidemic of mass psychogenic illness is to determine whether individuals will remain in a situation that is stressful to them. The view developed in this chapter suggests that those who are able to acknowledge that they have a problem, who can assess it as an externally caused problem, and who see medical experts as possible sources of amelioration, will leave the situation and seek aid before becoming a "case" in the dramatic way that will contribute to an epidemic. In that way, some individuals are probably "immune" to some of the forces which spread the epidemic (see Kerckhoff & Back, 1968).

Other factors besides personality undoubtedly affect the probability of an individual's becoming a "case," however. One of these is the position the individual occupies in the social setting. The position can be defined in both spatial and social terms. In spatial terms, the closer the individual is to the early cases, the more impressive they are likely to be. It is difficult to believe that the victim was "faking" or wasn't "really sick" if you were standing next to them when it happened. Also, if the explanation of the early cases of illness is made in terms of a toxic substance in the situation, anyone who was close to them can more easily believe that (s)he has been "infected."

Social closeness may be at least as important as spatial closeness. If one's best friend has been stricken, it is very difficult to argue that only "a bunch of crazy women" have been involved. Of course, if the epidemic continues over a period of days, as many have done, being personally close to an earlier victim is probably going to lead to visits to the hospital, discussions among mutual friends, and attempts to follow the progress of the case. In other words, the experience is going to be more intense and continuing than if one does not know the earlier victims personally. The importance of interpersonal relations in the transmission of mass psychogenic illness has been strongly suggested by the research to date (Kerckhoff & Back, 1968). The same social psychological perspective suggests, however, that *resistance* to the transmission may also depend on interpersonal relations. Being "immune" may thus depend not only on personal characteristics but also on one's social position. If none of one's acquaintances has been stricken, it is easier to reject the entire epidemic as something "out there" and not personally relevant. We might thus expect that nonvictims would be socially connected, just as we expect that victims would be socially connected.

WHAT IS A "CREDIBLE" EXPLANATION?

In a detailed case study of a case of mass psychogenic illness, Kurt Back and I (Kerckhoff & Back, 1968) argued that the explanation of the illness needed to be reasonable within the particular setting. In that case, the bite of a poisonous insect was a reasonable explanation as the epidemic occurred in June, in the South, where ticks and black widow spiders were known to be found. We pointed out that: "Although almost *any* kind of threat *might* have provided the basis for such an epidemic of symptoms (e.g., a vengeful spirit whose 'home' was disturbed by the building of the plant), the credibility of some kinds of threat will be greater than others [pp. 158-159]." In the context of the U.S.A. in the 1960s, that was a reasonable statement. Yet, we know that in another time and place the idea that illness might be caused by a vengeful spirit would be quite credible indeed.

An interesting example of the importance of the social and cultural context is provided by an epidemic in Singapore described by Chew, Phoon, and Mae-Lim (1976). They report a series of cases of illness (what they call "spells") in a TV assembly plant. Many of the cases involved violent behavior, screaming and fainting. In several cases, the women affected claimed to be possessed, to be "spirits" (the so-called "jin"). A local medicine man was brought in to help solve the mystery and to calm the workers. This kind of explanation of a series of cases and the action taken by management would hardly be credible in this country, but it was quite credible within that cultural context. The case in Singapore also indicates the importance of social relations in the affected population. Although only two-thirds of the workers in the plant were Malays (others were Chinese and Indians), all but one of the affected cases were Malays. Whether this concentration within the one cultural group was a function of shared cultural beliefs (which made the claims of being possessed more credible to the Malays than to the others) or due to closeness of personal relations (which made the early cases more significant to friends within the single cultural group) cannot be determined from the data available. But the seeming "immunity" of the non-Malays is very striking. Back and I noted a similar difference between black and white workers' susceptibility to the epidemic of insect bites, but in that case the social segregation of the two groups seemed to be the obvious explanation rather than differences in cultural beliefs.

It is interesting to note that in the Singapore case management called in the medicine man even though that action confirmed an explanation of the epidemic that management presumably did not accept. The action was taken to attempt to reassure the affected population that everything possible was being done to "cure" the illness. It seems likely that some epidemics closer to home have also involved such palliative actions by those outside the affected population. In order to reassure those affected, experts or managers may carry out procedures which they view as wholly unnecessary but which are used to prove the invalidity of a possible explanation of the illness or to prevent further cases attributed to that presumed cause. No generalization is possible, but it is at least reasonable to question whether such actions are always well-advised. Although a negative finding from such procedures may "prove" to the experts that the suggested explanation was incorrect, the very fact that the experts took that action tends to confirm the credibility of the explanation. More often, of course, the experts are at least originally unsure whether a truly toxic element is responsible for the illness, and thus they must carry out a series of procedures to search for such an element. Any such action tends to confirm that there *could* be toxic elements in the situation that could account for the illness. There seems no way, short of an initial confident diagnosis of mass psychogenic illness, to avoid presenting that possibility to the affected population. It may be wise, however, to avoid offering an unnecessarily long list of possible explanations, including explanations the experts view as not actually credible.

SEQUENCE AND SEVERITY

The view proposed here is based on the assumption that the initial cases in most epidemics of mass psychogenic illness are in the nature of random events, which then set off a kind of "chain reaction" of similar illnesses within the population. It may well be, of course, that the initial cases are due to the kinds of distress in the work situation that has been discussed earlier. These individuals may be under greater tension or be less able than others to cope with it. However, their illness may be due to entirely different causes. The first cases may actually be responding to a toxic element, a virus, food poisoning, etc. The important thing, though, is that their illness becomes *defined* as being due to something in the circumscribed setting and that others then associate any symptoms they may experience with this presumed source of threat. According to this view, the early cases provide evidence of a threat in the setting, and they thus both increase the level of tension experienced by others and provide a ready explanation for any symptoms others may develop.

The early cases are likely to be rather dramatic. The worker who feels ill and quietly goes to the infirmary or gets permission to leave work is not likely to stimulate the kinds of epidemics discussed here. Screaming, vomiting, fainting are the kinds of behavior that attract attention and suggest that a serious problem may be involved. Even then, however, the idea that there is an imminent threat will probably not immediately be widely accepted. But it is only necessary for one or two others to accept the idea, to respond overtly to the fear and to exhibit illness behavior, in order to increase the level of tension and raise the level of physical discomfort many others will experience. As the number of cases grows, the belief in the threat will increase, and overt responses to the threat will spread.

The responses to the threat will be of many different kinds. Those of us who study such epidemics focus on the "cases," those who are known to have become ill. Those individuals represent only one kind of response, however. Others may respond by withdrawing from the situation, by attempting to reinforce each other's skepticism about the threat, by mounting a corrective program (helping those stricken, complaining to management, etc.), or in some other way. It may well be, in fact, that many situations that might have become epidemics of mass psychogenic illness are reduced in severity by such responses.

In the epidemics that evolve sufficiently to be studied carefully there seems to be a pattern of legitimation involved that both increases the number of cases reported and reduces the level of severity of the symptoms in the later cases. As the number of cases increases, the explanation for them which evolves becomes increasingly convincing. At least it becomes increasingly difficult to deny that "something" is causing the illness. Because medical treatment is sought for those who become seriously ill, seeking medical attention is the reasonable thing for anyone to do if they believe they are becoming ill. In the later period of the epidemic, therefore, individuals in the population who experience even minor symptoms are likely to attribute them to the threatening element in the situation, and they are likely to seek medical aid before becoming seriously ill. The concerned response of the physicians tends to legitimize the reporting of even mild symptoms. As the number of cases grows, therefore, the severity of the symptoms reported to physicians will decline.

I noted earlier that when the experts take the illness behavior of the early cases very seriously and launch ameliorative actions and a search for causes, this tends to validate the belief in a threatening force and to increase the tension in the situation. Yet, one can argue that the continued involvement of experts can ultimately dampen the spread of the illness. By making it possible for those with mild symptoms to receive medical attention, physicians reduce the incidence of severe cases and lower the level of fear in the situation. Also, as the level of severity of the symptoms declines, even those who firmly believe in the reality of the threatening force are able to see the threat as one that is "going away." Whatever "it" was, it does not seem to be as bad as before. It may be true in some cases that "almost everyone" comes to believe that the illness was "only in their heads," and thus the epidemic ends in embarrassment. However, the general dampening effect due to expert attention and declining symptom severity seems a more reasonable explanation for the termination of most epidemics of mass psychogenic illness.

THE IMPORTANCE OF BEING FEMALE

It has been repeatedly noted that women are much more likely than men to be the victims in such epidemics. The evidence is too strong to deny the sex-linked nature of the phenomenon. Yet, this only poses the further question of why this should be so. No one is likely to accept the facile explanation that "women are just like that" without raising additional questions.

Colligan and Murphy (1979) have suggested that a consideration of the nature of sex roles in our society may help explain these observations. They note that "females have been encouraged to be sympathetic, loyal, submissive, and emotionally expressive". Therefore, these characteristics should lead women to be more active in an expressive social process such as found in these epidemics. They also point out that women in the labor force often face some serious role conflicts. Their work obligations may make it difficult for them to carry out their familial obligations in the way they would like to. Thus, women may experience more stress on the job than men because of these dual and sometimes conflicting obligations. The combination of greater tension and a more expressive personality can thus be viewed as a reasonable basis for explaining the preponderance of female victims of mass psychogenic illness.

Although I think that the Colligan and Murphy discussion has considerable merit, I would like to suggest another possible reason for the preponderance of female victims. When we review the work situations in which the best-documented cases of mass psychogenic illness have occurred, we find them to be strikingly similar. A room filled with rows of sewing machines operated by women, a series of long benches at which women assemble TV sets, a large room in which many women punch and verify computer data cards, and so on. These are highly regimented work situations in which the workers are all carrying out very boring, repetitive tasks. Hour after hour, they do the same thing over and over again. In most of the cases reported, the degree of regimentation seems to be very high and the workers are permitted only infrequent breaks between which they are restricted in both physical space and their ability to vary their activities. It would be difficult to find situations that would be more conducive to a sense of tension and frustration.

Without claiming to have detailed knowledge of the job distributions of men and women, I would like to suggest that women are more often found in such repetitive, boring, restrictive work settings than men are. It seems part of the folklore of industry that women are "good at" such tasks. They are supposed to tolerate the lack of large muscle movement and the constant repetition of small-part assembly tasks. Thus, we seldom find large numbers of men in such situations. It may be true that women tolerate those work conditions better than men, but it seems likely that those conditions are as much responsible for the high proportion of female victims of mass psychogenic illness as is the fact that women are expressive and suffer from role conflict.

Since employers are certainly interested in preventing the kind of epidemics discussed here, it is worth suggesting that one of the ways they may be prevented is to reduce the strain workers experience on the job. It is not reasonable to expect that women are so "different" that they can easily tolerate conditions which would be expected to drive men to rebellion. At least the work conditions are under the control of management, whereas role conflict and the presumed expressive nature of the female personality are not. Thus, although all of these factors may help explain the preponderance of femals victims, only one of them is controllable.

DIRECTIONS FOR RESEARCH

An important distinction between two kinds of cases of mass psychogenic illness is suggested by the differentiation made by Turner (1964) between convergence and contagion as the basis of similarity of behaviors in a collectivity. This distinction has been discussed by Milgram and Toch (1969), among others, and has more recently been included in a discussion of epidemics of mass psychogenic illness by Colligan and Murphy (1979). Turner was particularly concerned with the emergence of "new" or nontraditional forms of behavior which are not easily predicted in advance. The distinction he made was between (1) cases in which the similar behaviors are relatively independent individual responses to a similar stimulus (due to the "convergence" of similar persons) and (2) cases in which one person's behavior affects the tendency of another to behave similarly (through a "contagion" of the behavioral form). In fact, Turner argued that neither of these processes adequately explained many situations of collective behavior. Rather, he pointed to the importance of an "emergent norm" in such situations, a new definition that legitimized, often even encouraged, a form of behavior not previously deemed appropriate. He suggested that the behavior of the initial cases in many such situations did not so much "contage" as they helped to define what was a possible, useful, or satisfying way of coping with the situation. Others adopt such forms of behavior, then, because they provide an acceptable way of behaving in an ill-defined situation. Turner's (1964) emergent norm theory was thus intended to counteract what he and others viewed as the magical quality of the idea of behavioral contagion.

I would agree with Turner's basic idea that the spread of similar forms of behavior is seldom due to a simple process of contagion if contagion is viewed as direct imitation without purpose. I would prefer to use the term contagion to refer to a process by which one person's behavior elicits a similar response from another due to the meaning that the behavior has for the other person. In some cases, that meaning may be goal-directed, the behavior is seen as a good way to achieve a desirable goal. In the epidemics described here, the behavior of those who seek medical assistance when they begin to feel ill would be of that type. In other cases, the original behavior may serve to define the conditions in which the population finds itself. The early cases in an epidemic serve to define the setting as one that contains a dangerous element. If another person then becomes ill, it is not because (s)he "caught" the illness expressed by the other's behavior but because (s)he believed in the threat and experienced physiological symptoms of increased tension.

Even if we use the term "contagion" in this special or restricted way, however, it is clear that not all epidemics of mass psychogenic illness involve much contagion. In many epidemics, a number of cases occur almost simultaneously in response to a suddenly discovered (or suspected) toxic element. Someone smells a strange odor, and a number of people faint in response to this "gas." Others vacate the area, and the epidemic is over. It all happens at once and is quickly finished. It seems likely that such epidemics are the result of the convergence rather than the contagion process. By that is meant that the individuals who become victims are responding more in parallel than in sequence, their personal qualities are much more fully responsible for the behavior than are the kinds of worker-worker and expert-worker relations that have been discussed here.

Undoubtedly there are no really pure examples of convergence or contagion. In the present discussion, for instance, we assume that one or more early cases initiate the contagion process. In most epidemics, it seems reasonable to view several of the early cases as relatively individual events involving little inter-individual influence. And, even in the most rapid response epidemics, all cases do not occur simultaneously and thus it is at least possible that later cases have been influenced by the existence of earlier cases. But, the two kinds of situations are very different, and it does not seem reasonable to expect the same generalizations to apply to both. At the time we were engaged in the analysis of *The June Bug* data, we investigated another epidemic that followed a pattern expected from the convergence view of these epidemics (Roth, 1965). We were unable to find any pattern of interpersonal relations among the victims, even though such a pattern was a prominent feature of *The June Bug* analysis. Personality characteristics seem to be a much more central feature in cases where the convergence perspective is appropriate. Future research should attempt to differentiate between these two kinds of epidemics and to assess the extent to which different kinds of explanation are needed for them. For instance, is the convergence type more likely to occur in populations having a high proportion of a particular kind of personality, and/or an over-all higher level of tension? Is it possible that the contagion type can occur in any population having a basic set of structural characteristics, only requiring a single or small number of initial cases which are defined appropriately to heighten the level of fear?

This last question leads to another set of issues that the earlier discussion has made salient. Basically, these are issues dealing with the structure of the situations in which the epidemics occur. Are some structures especially likely to generate tension as I have suggested earlier? If so, are there ways in which the tension may be relieved so as to avoid the kinds of epidemics we have observed? Although this is presented as an important research question, it is obviously an important management question as well. It may be possible to observe various kinds of structured situations and to assess their tension-producing potential without waiting for one of the explosive events to occur. This would require a better set of measures than I am aware exists for indexing collective tension, but it would be well worth the investment to attempt to develop such measures.

The situation changes, of course, when the early cases occur. A set of actions, both official and unofficial, are initiated at that time. We are beginning to have available information about a sufficient number of cases that it ought to be possible to begin to estimate what kinds of actions tend to increase and what kinds tend to decrease the probability of additional cases of illness. With regard to official actions, the earlier discussion has suggested two opposite patterns. It has suggested that the intense and serious involvement of experts in the situation may serve to increase the credibility of a serious threat thereby raising the level of tension and increasing the probability of additional cases. Yet, it has also suggested that the involvement of experts should ultimately lead to a reduction in the incidence of cases with severe symptoms and will help convince even firm believers that, whatever the problem might have been, it has been removed. It is possible, of course, that both of these statements are correct. An initial increase in tension may be the price that must be paid for the ultimate solution to the problem.

It seems unlikely, in any event, that the involvement of experts can be avoided. Unless we are able to specify for industrial and school physicians what symptoms are found only in cases of mass psychogenic illness, these epidemics will continue to require the serious and intensive involvement of medical and other kinds of experts. In one of the more widely-read discussions of such epidemics (Roueché, 1978) it is suggested that if the attending physician or other authority figure takes a firm stand, (s)he can stop an epidemic before it becomes more serious. If (s)he simply states authoritatively that there is no toxic element in the situation and that the symptoms are due to tension and fear, it is possible to return the situation to normal in short order. Yet, in that same discussion, the physician who accomplished this transformation is quoted as having said, ''But, thank God, I was lucky.'' It isn't likely that many experts will take such a chance, and it is not likely that many people would want them to do so. Yet, more careful study of the immediate symptoms and the pattern of spread of the early cases may make it possible to make such decisions with greater confidence. One way to do this would be to record as carefully as possible the sequence of events in epidemics that do occur, even if this must be done retrospectively. A systematic review of a series of such records could then make it possible to determine when one could make a confident determination that the epidemic is one of mass psychogenic illness and not some other form.

It should be emphasized in closing, though, that in order to approach these epidemics in a way that is most promising for diagnosis and definitive action, something besides the usual medical perspective must be adopted. In the present discussion I have adopted a terminology I assume most physicians would have used, at least in that I have referred to individual ''cases.'' The ''case'' in this and most discussions is an individual who comes to the attention of a physician, is examined and, possibly, given some form of treatment. To make the kinds of analysis suggested here, and to be able to take the kind of definitive action that might effectively dampen an epidemic with strong psychogenic components, it is necessary to see the whole *series* of individual cases as the focus of analysis. The clues to the overall diagnosis will be found as much in the pattern of cases and in the overall distribution of symptoms as it will in the individual cases. The physician may not always be able to ''step back'' and take such a broader view—if nothing else, he may be too busy coping with the flood of individual cases—but someone needs to do it.

REFERENCES

Bart, P. B. Social structure and vocabularies of discomfort: What happened to female hysteria? *Journal of Health and Social Behavior*, 1968, *9*, 188–193.

Chew, P. K., Phoon, W. H., & Mae-Lim, H. A. Epidemic hysteria among some female factory workers in Singapore. *Singapore Medical Journal*, 1976, *17*, 10–15.

Colligan, M. J., & Murphy, L. R. Mass psychogenic illness in organizations: An overview. *Journal of Occupational Psychology*, 1979, *52*, 77–90.

Kerckhoff, A. C., & Back, K. W. *The June bug: A study of hysterical contagion*. Englewood, N. J.: Prentice-Hall, 1968.

Levine, R. J., Sexton, D. J., Romm, F. J., Wood, B. T., & Kaiser, J. Outbreak of psychosomatic illness at a rural elementary school. *The Lancet*, 1974, *2*, 1500–1503.

Markush, R. E. Mental epidemics: A review of the old to prepare for the new. *Public Health Reveiws*, 1973, *2*, 353–442.

Milgram, S., & Toch, H. Collective behavior; crowds and social movements. In G. Lindzey & E. Aronson (Eds.), *The handbook of social psychology, Volume 4*. Reading, Mass.: Addison-Wesley, 1969.

Roth, R. H., Jr. *A study of hysterical contagion*. Unpublished M. A. Thesis, Duke University, 1965.

Roueché, B. Annals of medicine: Sandy. *The New Yorker*, 1978, *21*, 63–70.

Sirois, F. Epidemic hysteria. *Acta Psychiatrica Scandinavica, Suplementum*, 1974, *252*, 7–46.

Turner, R. H. Collective behavior. In R. E. L. Faris (Ed.), *Handbook of modern sociology*. Chicago: Rand-McNally, 1964.

13 Perspectives on Epidemic Hysteria

François Sirois
Centre Hospitalier de l'Université Laval

ABSTRACT

A brief historical survey of collective hysteria gives perspective to the topic. Through the variance of medical and cultural labels a core clinical outbreak is sifted out of the recent reports assembling the descriptive aspects of this complex phenomenon. Epidemiological variables are reviewed with special emphasis on outbreaks occuring in occupational settings. The epidemic process is studied from the clustering of outbreak curves leading to the hypothesis of three contributing elements: the prodroma, the epidemic moment, the rebound; each one with its own elements and method of study. The making of the outbreak is linked to group factors as they relate to psychodynamic psychology.

Fads, crazes and fashions have spread in waves over mankind since time immemorial. Nowadays the global village has speeded up, highlighting the scope of massive behavior dissemination. In the maze of various crowd behaviors such as riots and panics, and larger social trends or selected cultural patterns such as revitalization movements, there is room for socially shared psychopathology.

Until recently, collective mental disorders received more anecdotal average coverage than systematic inquiry. In the no man's land between psychiatry and various social sciences, the more salient are the epidemic aspects of hysteria, suicide, and paranoid psychoses. The recent horror of the Jonestown Suicide in Guyana echoes other less spectacular trends since Durkheim's classic study of suicide. Likewise modern large scale persecutions hide the background of scores of clinical studies on smaller units of group paranoia, "Folie à Deux" and familial delusions being the best documented.

We present here a brief historical survey of epidemic hysteria, a comprehensive review of the world literature with special reference to outbreaks in industrial settings and a personal view of the epidemic process.

HISTORICAL SURVEY

A Focus for Folklore

Various collective "hysteriform" manifestations have been known from antiquity through the dyonisiac tradition. The *dancing manias* of middle ages appeared in Aachen in 1374 some time after the Black Plague epidemic of the XIV century. They spread rapidly along the Rhine basin and through pilgrimage routes of German stock Hungarians to St-Vitus sanctuaries. Dances were mostly from lower social groups of peasants, small artisans, young people of both sexes, and idlers. Oft naked, they held or clapped their hands, went into ecstatic frenzy, and threw themselves to the ground with paroxysmal convulsive movements and twitchings. Groups of a few hundred people were involved. Outbreaks were recorded sporadically for about two centuries and were sketched by Brueghel in 1564. Paracelsus in his book of 1531 on invisible diseases discussed the so-called St-Vitus dance attributing the sickness to the irrational power of imagination and belief. A local variant was the phenomenon of Tarantism in Italy. The attacks were mostly noted in summertime among peasant women who would dance frantically and present peculiar behaviors: animal-like squeaking, obscene shouting, a desire to be tossed in the air. Dancing with music until exhaustion was considered a remedy for the tarantula bite (Hecker, 1844; Martin, 1923, Rosen, 1962).

A second wave of historically visible outbreaks were the epidemics of *demonopathy* in the convents and in rural areas of Southern France. Calmeil described a dozen of these occurring between 1550 and 1650 at the height of religious wars (Calmeil, 1845). Widely known are the ones in Aix in 1609 and Loudun in 1632. The climate of the Catholic Inquisition, framed by the "Malleus Maleficarum" which dealt with the legal examination and sentencing of witches, arose at a time of strains and tearing of the social tissue with the exploration of new worlds, the uprising of free thinking and the rupture of the religious order. Most of the episodes were reported along the geographical clash of the Huguenot and Catholic world in France. After a short plague outbreak in 1632 in Loudun, a demonic outbreak flared up in a convent. The chaplain was accused, his morality at stake. The girls were sick but the man was burned. De Certeau discussed the delicate intrication of libidinal, religious and political transgression (De Certeau, 1970). Although some of these outbreaks could have been attributed to avitaminosis or ergot poisoning, most cases leave little doubt about their hysterical nature. The symptoms (globus, cries, convulsions, abnormal and symbolic movements) do not hide completely the obvious underlying sexual and aggressive conflict in the detailed chronicles. Women were almost exclusively involved in similar rural outbreaks (including the regions of Savoie, 1574; Languedoc, 1577; Pyrenee, 1600; Nimes, 1632) which were reported sporadically up to the XIX century (Calmeil, 1845).

The legacy of these phenomena can be traced in a third constellation of outbreaks which appeared again in a religious climate of dissention and sectarism. We think of the Jansenist convulsionaries of Paris in the mid-eighteenth century, the Camisards revolt in the Cevennes of France. Following the exile of Huguenots to England after the revocation of religious tolerance in 1685, group excitation and enthusiasm have been associated with a variety of religious movements notably the early Quakers with the additional feature of running naked. Religious meetings were often marked with group ecstatic phenomena and prophetic trance. The Shaking Quakers later fled to America. Similar religious revivals developed especially in the Appalachian frontier during camp meetings aroung 1800, where groups of Shakers, Jerkers, and Barkers became familiar. Later in Britain, Methodist revivals in Cornwall and Belfast witnessed similar phenomena. Other manifestations were known to show related characteristics within Judaism (e.g., the Zar Cult in Ethiopia), within Islam (with the Soufi tradition in Persia disseminating to Malaysia and Indonesia), or within maraboutism in the Maghreb and Twirling Dervisches in Asia Minor (Cornish, 1814; Robertson, 1806; Rosen, 1962).

In modern times beyond the West, a well known outbreak was reported from Madagascar in 1863 with collective dancing and convulsion following the cruel reign of queen Ranovalona I on rumors of her comeback. Japan has known similar outbreaks at the close of the Tokugawa era. The best known was in 1867 just before the Restoration. After spells of uprisings and economic crises, crowds would chaotically dance and sing in an unusual manner. In Brazil, when slavery was suppressed around 1896, more politically cohesive outbreaks appeared around A. Conselheiro (Davidson 1867; Nina-Rodrigues, 1901; Norman, 1945).

In Europe, from the Reform to the Industrial Revolution, the setting up of the General Hospital and related institutions, specially in the seventeenth century, shifted the social handling of misfits from exile to internment. Outbreaks were then reported from prisons, orphanages, hospitals and later in schools where the bulk of outbreaks in the XIX century was recorded (Bricheteau, 1863; Calmeil, 1845; Chambrelin, 1849). With the industrial revolution, a shifting economic order brought with it epidemic hysteria in mills and factories. The first modern outbreak was reported from a Lancashire cotton mill in 1787, in the Gentleman's magazine. It must be remembered that Cartwright invented the power loom in 1786 following a fast mechanical revolution in the textile industry that overthrew the secular handworking order with many consequences for the ethics of labor.

The meaningless monotonous working hours, the abused young people and the massive urbanisation generated compensatory needs for excitement and evasion which were sought elsewhere.

"Oppression, insecurity and the influence of a very rude priestcraft are the powerful causes which operated on the Germans and Italians of the middle ages"; these remarks made by Hecker (1844) in his study of the Dancing Manias summarize the climate under which most of these socially shared types of psychopathology made their appearance. Religious intolerance, political unrest, economic insecurity, cultural clashes, changing social orders, natural disasters or devastating diseases were historically an ideal ground for the breeding of similar phenomena. Many authors noted that poorly educated, superstition-prone impoverished classes had always been more involved. Most of the outbreaks are obviously much more complex and should be interpreted along many lines of thought. It is not our purpose and beyond our competence to prove that they really were. They have been presented to gain more perspective into the phenomenon of epidemic hysteria. The term is used to underline the contagious, unorganized, uncontrolled nature of the process, which is recognized as spontaneous, unusual, odd and which is given neither support nor suppression by the whole community. Therefore we shall exclude ritualized and institutionalized contagious behaviors related to integrated religious or civic customs. Some of the outbreaks related above were later incorporated into the folklore. At this phase, they no longer belong to our subject. They pertain more to social character traits than to social symptomatic acts.

CLINICAL ASPECTS

A Century of Scattered Observations

Since 1800, outbreaks of epidemic hysteria became increasingly reported through the technical literature, most of them occurring outside the religious climate of earlier times, and more amenable to clinical analysis. Hardly accessible and descriptively uneven, over a dozen are to be found in the early French and German papers. Over the last 100 years, there has been a striking variability in the frequency of reporting. During the classical period from 1870 to 1913, 43 papers analysed outbreaks, half of them in German, the rest in Italian, Russian, and French journals. Interest in both hysteria and epidemics was high at the time of Charcot and the expansion of the germ theory. During the Wars, 1914–1943, epidemic hysteria became a rarity in the literature. Much attention was given to intrapsychic dynamics or neuropsychiatric problems. Only eight reports were located. Since 1943, there has been a steady rise in reported outbreaks following the first contemporary description by Schuler and Parenton (1943). The renewal of interest has been mediated by sociology. Fifty recent articles are to be found, over 80% in the English speaking journals.

To map out the clinical picture of epidemic hysteria, a comprehensive survey of the world literature of the last 110 years (1870–1980) was carried out of 88 reports, the last 13 being unaccessible (Sirois, 1974).

(I) Persons, place and time. Half of the outbreaks were reported from schools, a quarter from towns and villages, 17% from factories, the rest occurring in various institutions. Women were almost exclusively involved (80%), especially for closed settings, but less so for those appearing in towns. They were young, mostly adolescent girls, in 60% of cases. The number of persons affected in an outbreak varies from a few to over hundreds in exceptional cases. There is an equal distribution for small groups of less than ten, class size groups of 10 to 30 and larger groups. The outbreaks were either shortlived (1 day) or could last up to a few months but usually ranged between three days to two weeks.

(II) Symptoms. The symptomatic pattern varies according to the period of reporting. Outbreaks prior to 1914, were made up mostly of globus, laryngismus, or choreiform movements. After 1914, there is more fainting, nausea, abdominal malaise, headaches or dizziness. Convulsive movements, tremors, hyperventilation are evenly represented. These behaviors are generally brief, lasting less than one day. Relapses in the same person are not the rule except for index or early cases. A second wave is usually least affected and not relapsing. A

TABLE 13.1
Leading Variables of 88 Reviewed Outbreaks

Location		*Sex*	
Schools	41	Female	70
Towns and villages	20	Female and male	13
Factories	15	Male	4
Institutions	5	Unknown	1
Hospitals	3		
Others	3		
Unknown	1		

Number of people Involved		*Duration*		*Age*	
Less than 10:	27	<3 days:	16	<20:	51
10–30:	28	3–14 days:	34	20–40:	22
more than 30:	28	15–30 days:	11	all ages:	9
unknown:	5	>30 days:	17	unknown:	6
		unknown:	9		

secondary attack rate in households is absent. The outbreaks flare up after a notoriously short incubation period, with plenty of symptoms, little morbidity and a scarcity of signs. Associated laboratory measures are usually normal except for those linked with hyperventilation. Associated excitement or fear is usual, and does not affect adults.

(III) Individual features of affected. The usual psychological tests do not yield convincing evidence that the majority of affected persons will display psychopathological traits or below average I.Q's. The more constant finding is higher N scores on the Eysenck Personality Inventory. One opinion is that epidemic hysteria occurs in a group including *a large majority of normal per*sons where the threshold effect for suggestibility plays a major part (This is due to the regressive effect of crowd behavior on intellectual and emotional processes), and *a small minority of unstable or emotionally* labile individuals which can often be found to grevitate around index cases. Moss (1966) noted that the epidemic spread down the age scale. An index case was identified in over 50% of outbreaks.

(IV) Group characteristics. The usual case control approach appears to yield less than the study of the total group as a unit and the global epidemiological configuration. The study within the affected group is usually conducted with a sociometric approach to establish patterns of spread, mostly by sight and sound or physical proximity, otherwise by affiliations. The relevance of interpersonal networks in the diffusion of new behavior in situations of uncertainty has been established (Coleman, Katz & Menzie, 1957). Two points of view are expressed here. Kerckhoff, Back, and Miller, (1968) noted the early dissemination between social isolates, less subject to initial pressure toward expected conformity and more vulnerable to contagious behaviors. Other reports showed that outbreaks disseminated from the beginning among strong interpersonal networks. Reports by Teoh and Tan (1972) and Benaim, Horder, and Anderson (1974) show the complex configuration of insiders and outsiders, suiting both points of view according to the perspective. H. B. M. Murphy (personal communication) suggested that the persons who stand out in the group—isolates, leaders, new comers—are the likely triggers.

Another group situation is the rapport of the affected group with the immediate environment. Isolation and cohesiveness are group characteristics believed to be important in the developing conditions of an outbreak. Isolation of a group prevents the verification of perceived threats with the outside. Less isolated groups such as tribes, hospital personnel or working staff are probably made more cohesive by cultural or occupational conflicts (Stahl & Lebedun, 1974). The same holds for alleged external threats in outbreaks occurring in towns fostering togetherness in open groups. Some outbreaks in school settings occurring at the beginning of the academic year could be seen as ''rite de passage,'' to bind anxiety linked with the formation of the group and act as witness of collective cohesiveness. The same could apply to excited marching bands' outbreaks (Levine, 1977). We will expand later on the relative weight of precipitating events or rumors and the larger predisposing factors which breed the outbreak. The social issues are usually elaborated upon where cultural differences make them obvious, otherwise often neglected (Armstrong & Patterson, 1975; Eastwell, 1979). The precipitating factor in itself is a rather innocuous event.

(V) Analysis of outbreaks by settings. Out of the bulk of varied reports, a closer picture is seen through a comparison of outbreaks according to where they happen. Among school outbreaks, the ratio of affected females relative to males is four to one. The age is below twenty. Duration is bimodal, either short or long. The number of people involved varies. Before 1943, class size groups made up most reports. Later, larger groups were involved as if there were a change in the reference group; stronger larger units like marching bands or cafeteria groups developed. The attack rate is usually around 30% for both group sizes. Factory outbreaks affect females only in the 20–40 age range. Duration is short, 30% below 3 days, 70% up to 2 weeks. A large number of people are affected. The attack rate varies inversely to group size, smaller in large groups, larger in small groups. Town outbreaks involve both males and females of all ages. Duration and the number affected are variable, and the attack rate is unknown because the denominator cannot be estimated. Institutional outbreaks are seen mostly in a small group of young post-adolescent segregated females. Duration is usually long and the attack rate is around 15%. (See Table 13.2.)

TABLE 13.2
Analysis of Outbreaks by Setting

	Factory	*School*	*Towns*	*Institution*
Sex	F	80% F	M+F	F
Age	20–40	<20	Bimodal (20+all ages)	Young Post-adolescent
Duration	short	Bimodal	Bimodal	Long
Number of people involved	large	Class size <1943 Large group >1943	Variable	Small
Attack rate	Inversely variable with group size. (10% = large group) (30–50% = small group)	30%	?	~15%

(VI) Aspects of outbreaks in occupational settings. A sharper focus on outbreaks in this category shows two types of situations. The first is the health related occupational outbreak. A report from Japan illustrated a breakdown of the power of the nursing role in the changing social system of a laprosarium from a closed, medieval outcast society to a modern hospital. It was characterized by alienation and hopelessness (Ikeda, 1966). More acute fear-provoking situations reported by McEvedy and Beard (1970a, 1970b) showed a helpless and captive staff shift to a situation where they had to be taken care of as patients instead of assuming their duty of caring for patients.

The second type of situation occurs in industrial settings where conditions of productivity are at the forefront. The outbreaks generate a decrease of work output with disruptive consequences. The prototype of such outbreaks was reported in 1787 in *Gentleman's Magazine* by a Dr. St. Clare at a cotton manufactory:

> A girl, on the 15th of the month, put a mouse into the breast of an other girl who had great dread of mice. She was immediately thrown into a fit and continued in it with the most violent convulsions for 24 hours. On the following day, three more girls were seized in the same manner, and the day after, six more. By this time the alarm was so great, that the whole work in which between two and 300 were employed was totally stopped, and an idea prevailed that a particular disease by a bag of cotton opened in the house had been introduced. On the 18th, the Doctor was sent, before he arrived three more were seized, and during the night and the morning on the 19th, eleven more, making in all 24. Of these, 21 were young women, two were girls of about ten, and one man, who had been much fatigued with holding the girls. Three lived about two miles from the place where the disorder first broke out, and three at another factory, about five miles distant, and two more were infected entirely from report, not having seen the other patients; but like them and the rest of the country strongly impressed with the idea of the plague being caught from the cotton. The symptoms were anxiety, strangulation, strong convulsions, so violent as to last from a quarter of an hour to twenty-four hours, and to require four or five persons to prevent the patients from tearing their hair and dashing the head against the floor of the walls. By electric shocks the patients were relieved without exceptions. As soon as the patients and country were assured that the complaints was merely nervous, easily cured, and not introduced by the cotton, no fresh person was affected. To dissipate their apprehension still further, the best effects were obtained by causing them to take a cheerfull glass and join in a dance. The next day they were all at work [p. 268].

This pioneer description gives an accurate picture of the typical outbreak. It is explosive, bimodal with peak in day 3 and 5, lasts about a week, with an attack rate of 10%. The rumor plays a central role. The 15 outbreaks reported since then shows that symptoms can fluctuate according to the type of work and the current belief about the noxious agent as illustrated by Table 13.3. An index case is reported in 10 outbreaks. The attack rate seems inversely related to the size of the

TABLE 13.3
Factory Outbreaks (1787–1979)

Year	*Country*	*Type of Plant*	*Attack Rate*	*Symptoms*	*Duration*
1787	England (Gentleman's Magazine)	Cotton	10% (24/200–300)	Convulsions	6d.
1875	Germany (Seelig Muller)	Farming	? (9/?)	Convulsion, globus	10d.
1882	France (Bouzol)	Silk	? (12/?)	Abnormal movements	3w.
1891	Russia (Schatalow)	Lace	? (20/?)	?	5d.
1914	Russia (Bechtereff)	Rubber	? (90/?)	?	1w.
1941	Italy (Franchini)	Soap	? (16/?)	Convulsion, tremor	1d.
1956	Italy (Parigi)	Goldwork	5% (30/700)	Hyperventilation, tremor	2d.
1956	Australia (Bell)	Textile	33% (55/166)	Dizziness,	12d.
1962	USA (Champion)	Textile	7% (62/965)	Fainting, paresthesia	1w.
1972	USA (Stahl)	Data center	50% (30/60)	Dizziness, fainting	2d.
1972	Malaysia (Chew)	Plastic toy	2% (5/250)	Crying-screaming	4d.
1973	Malaysia (Chew)	Television set	8% (84/1000)	Crying-screaming	8d.
1973	Malaysia (Chew)	Fish food	10% (34/350)	Crying-screaming	13d.
1976	England (Maguire)	Ceramics	? (10/?)	Dermatitis	?
1976	England (Maguire)	Textile	? (17/?)	Dermatitis	?
1978	England (Cunliffe)	Machinery	? (8/?)	Dermatitis	2–3d.

group, about 8 to 10% in large groups and from 30 to 50% in small groups (Bekhtereff, 1914; Franchini, 1947; Schatalow, 1891; Seeligmüller, 1876).

The precipitating factors are often found inside the factory either as an implicit psychological condition like job dissatisfaction, or a rather innocuous material event: installation of a ventilator (Parigi & Giagiotti, 1956), fumigation of the plant (Bell & Jones, 1958; Champion, Taylor, Joseph, & Hedden, 1963), oil change (Cunliffe, 1978). Otherwise it is to be found outside the industrial setting: in society (Chew, 1978), with possible cultural clashes between multiracial staff, or in individuals as in Bouzol's (1884) report where the manufacture merely acts as a reunion point like school. Little attention is given to larger predisposing factors which would explain the general atmosphere of work. The impression is that the precipitating factors act as a screen for disturbing factors.

Some underlying fantasies revealing these factors can be inferred from reports, especially the study of rumours. Maguire (1978) reports a central couple fantasy, with male and female working together; it could be either a birth wish, creating new jobs, or a coital wish, copulation being better than ceramics. A gas leak fantasy is often found. Workers wishing or fearing to be put to sleep; often an expression of bored, sleepy workers as victims of taylorized duties, escaping through day dreaming. A contamination fantasy is found in the fumigation cases. As if the workers were insect-like, working hard, blindy, at the limit of their capacity in a beehive manner. We find here all three variants of the basic assumption of working groups described by Bion (1959).

Another aspect is the basic unit of the working group. It is often found that such outbreaks spread inside working shifts mainly, the rotation of staff disbanding the disturbed group, providing a frame for early focused intervention preventing the total disruption of the plant. The risk of endemicity for outbreaks in occupational settings appeared low. That we should be reminded of this possibility like the 1919–1920 ripe olive scare that threatened the whole canning industry in the USA following outbreaks of botulism. This risk would probably be higher with any reality-based danger but small once it is a fantasy-based danger.

The main task is an accurate evaluation. The major differential diagnoses are: 1. heat stroke, with expected incidence of 2/1000, associated syncope with bradycardia and a fall in blood pressure, 2. food poisoning, 3. gas intoxication, 4. allergens, 5. infections, usually with longer incubation period and secondary attack rates in household.

(VII) Treatment. This brings us to the issue of treatment, the basic principles being known since earlier times as Bozzolo put it in 1894: "L'effetto dell' isolemento fu splendido." The breaking of the group, the closing of the plant in severe cases, the isolation of members remain the most effective way to abort an outbreak. Coercive measures, like public confrontation and exhortation are usually useless and aggravate the outbreak. Adequate information, attention to both reality where reassurance is needed and the fantasy behind the episode especially dealing with the rumour should provide some resolution of the crisis. Counter lying to repress or evade the element of truth behind the deceptive outbreak might be a dangerous weapon with short term results but with long term chances of recurrence. Once collective psychopathology is identified the issue of reassurance brings about the manner of disclosing the situation. To announce that it is *only* emotional—hence groundless—might have a positive effect if fear is involved in a tension reducing outbreak. But if evasion through excitation is searched for, or more so if protest is the issue under the episode it might spur other alternative disruptive episodes if the claim behind is to be ignored.

(VIII) Course and occurrence. Spreading factors promote the lasting or the rebounding of the episodes. These include mass media, physical and visual proximity, excitement, reuniting the group, and the persistance of rumors in long lasting outbreaks (Kagwa, 1964; Rankin & Philip, 1963). It was often found that the reservoir of susceptible persons was much larger than originally suspected. This meant that important and more general aspects necessary for the understanding of the situation were missed and the original group was only instrumental in the outbreak. Other long lasting episodes (Knight, Friedman & Sulianti, 1965; McEvedy & Beard, 1970) were brought about by some inability or impossibility to reach or modify the conflicting situations.

Two follow-up studies of detected outbreaks have been conducted. They explored the long range fate of affected individuals but did not give further information about global aspects of outbreaks such as absence of recurrence or conditions involved in its disappearing. The first one was conducted by Rawnsley and Loudon (1964) 25 years after the outbreaks of 1937 in Tristan da Cunha. They showed that women then involved would later be characterized by frequent medical consultations, marriage to leaders of the community and frequent headaches. McEvedy and Beard (1973), using a case control approach studied the 1955 Royal Free Epidemic 13 years later. They showed that affected individuals had more neurotic scores on the Eysenck Inventory, poorer health, more hospital admissions, and lower fertility rates. They also gave evidence of simulation in a few cases and a high rate of sick leave before the epidemic in a group of affected. The two studies favor a differential threshold for hysterical behaviors which would be activated and lowered during the epidemic moment. The retrospective differentiation of pathological subjects from normal ones to the affected seem easier with the passage of time.

The incidence of hysterical epidemic outbreaks is not well known. A pilot study conducted (Sirois, 1975) in the schools of the province of Quebec in 1973, estimated the attack rate to be around 1 per 1000 school settings per year.

Epidemic hysteria has an endemic propensity. This holds for historical periods, special social settings such as schools where it is especially frequent, and finally for certain geographical areas such as Malaysia (Teoh & Yeoh, 1973, and Teoh, Soewondo & Sidartha, 1975), and Africa. Two examples will serve to illustrate the discussion about endemicity.

McEvedy and Beard (1970) reviewed some epidemics of pseudo-poliomyelitis. Ten were reported between 1952 and 1958. In the early 1950s a major epidemic of poliomyelitis in the U.S.A. caused about 50,000 casualties. This was the climax of a new danger which justly frightened the population just as the Middle Ages were struck by the Black Plague. Major efforts led to massive vaccination since 1955 which cured both the problem and the anxiety linked with it. Most of the so-called benign encephalomyelitis outbreaks reviewed by the authors occurred in the nursing staff in close contact both with the medical preoccupations and worries and a potentially large contaminating milieu. They were put in a threatening, contradictory and unpredictable situation. Likewise, endemic outbreaks in Malaysia were detected in schools in 1970–71, after the 1969 troubles and in a context of administrative and educational changes (promoting the Malay population still disadvantaged as compared to the Chinese minority). Major floods were identified as the precipitating factors.

Our hypothesis with regards to endemicity is a psychodynamic one. The incubating situation is provided by a social context of change where uncertainty is high and tensions manifest themselves at the weakest point. The most important aspect is that tensions are unspoken and untold. Epidemic hysteria is thus used by society because it enables concerned people to displace or camouflage the real conflicts. Therefore the outbreaks become the focus where both the individual and social conflicts are superimposed upon each other and cannot be distinguished insofar as both are repressed out of consciousness and become identical in the Unconscious. A precipitating event of some sort mobilizes individual fantasies which are shared. The fear and anxiety then produced are equaled at a social level with a hardly perceived context of social uncertainty for which the outbreaks become a latent solution. There is a mirroring of the social conflict in the reality and the individual conflict in the fantasy, and the outbreak acquires the value of an unconscious symbolic solution for the social stiuation. Teoh's studies suggest that the socially repressed feeling was that traditional schools were better. Repressed because unacceptable in the large modernizing school reform at the time of the outbreaks. Yet the presence and successful activity of traditional healers during the episodes seems to attest the point.

THE EPIDEMIC PROCESS

A Personal View

Epidemic hysteria can be studied obviously from various theoretical frameworks according to the eye of the investigator. The epidemiologist physician will produce evidence for exclusion of organic outbreaks with a careful analysis of the epidemiological configuration. The psychologist will be inclined to compare affected versus nonaffected people in a case-control approach. The sociologist emphasizes the inside group study with sociometric techniques. The anthropologist focuses on the position of the group with its environment. The business consultant will advise on personnel management. The psychiatrist might stress the unconscious fantasies vehiculed in the outbreak, the symptomatic pattern of anxiety, the psychological defenses.

A factor hindering the study of outbreaks is to decide how to approach, observe, register and analyse such phenomena (Friedman, 1967). There is always something that falls in between the screening lens of investigators. The loosely defined boundary of epidemic hysteria falls on a spectrum between organic diseases on one side and social movements on the other. Misclassification occurring on both sides as hysteria camouflages itself within various elusive domains or covered social values according to the Zeitgeist. Late in the 17th century, Sydenham emphasized it could mimic almost all known physical diseases.

Epidemic Types

Despite all these difficulties, a further step in the study of the epidemic process can be achieved by the inspection of epidemic curves. According to a combination of variables of time, place and persons, the outbreaks appear to have various profiles, tentatively classified in five types.

The first one is the sudden onset explosive type. Symptoms appear rapidly, involving at once many people and are usually short lived. It appears to be a tension discharge phenomenon in a closed group. The benign and self-limited presentation is often given a superficial description which makes it difficult to both trace a triggering factor and propose a careful explanation. At times, it is a violent event in a school, new machinery installed in a workshop, a person falling ill in an institution, the beginning of a school year, a zoonosis in a rural area. This type has no specific location but hits small groups of young people (Lyons & Potter, 1970; Pfeiffer, 1964). Their abortive and spectacular nature makes them appear in newspapers headlines. Only the epidemic moment is visible.

The second type is explosive with an identifiable prodromal stage. Isolated cases are first detected. There is a gradual build-up of tension and the explosive outbreak follows. Index cases are probably perceived either as threatening to the group or used to carry and express some other shared conflicts. This lasts about one to two weeks according to the prodrome. It is a variant of the first type (Champion et al., 1963; Regnard & Simon, 1876; Schuler & Parenton, 1943). Its interest lies in the recognition that something is going on before the outbreak.

The third type is the cumulative outbreak. Less than 10 persons are involved and the transmission of symptoms occurs over a longer period of time—two weeks to a month—as a slow chain reaction. This type is found in closed institutional settings (Armainguad, 1879; Michaux, 1952; Taylor & Hunter, 1958). Because of the slow process these outbreaks lend thenselves to more detailed inspection of underlying fantasies but can go undetected. The prodrome is identical with the epidemic moment.

The fourth type is the rebound outbreak. A handful of cases appear rapidly and in the following days a more important second wave of people become symptomatic. The outbreaks last one to two months and involve larger groups (over 30). The rebound phenomenon often appears when the group is reunited. Absenteeism after the incident is important. In other cases, the rebound occurs as multiple successive epidemic relapses over a few months (Colin, 1880; Knight et al., 1965; Rawnsley & Loudon, 1964). This type is an ideal one for a detailed epidemiological study (Bouzol, 1884; Knight et al., 1965; Moss & McEvedy, 1966) since it deals with large numbers, a good span of time, detectable peaks, the possibility of control cases and the higher probability of detecting index cases. In a way it is epidemic hysteria in its most complete manifestation.

The last type is the large diffuse outbreak. It appears in communities, rural areas or towns. It is not restricted to specific groups. A fairly large number of people of both sexes and all ages could be involved. This type is somewhat more distinct from the others. In various instances, it appeared as a laughing and running impulse in African communities (Rankin & Philip, 1963), a Koro epidemic (Ngui, 1969), a panic stricken town overwhelmed with false rumors and beliefs (Johnson, 1945). The prodromal stage is always obscure, but the rebound stage is often spectacular. It is the most difficult to investigate.

To understand these various profiles, it is our impression that they all presented various aspects of an underlying process which shows up differently according to the moment of the epidemic they emphasize and the speed with which the process is unfolded.

The Epidemic Process

The outbreak is made out of three elements: Prodrome, Epidemic Moment and Rebound. The epidemic moment is always manifest in all five types. Only in slowly developing episodes, is it possible to study the prodrome; only in large or long lasting episodes it is possible to study the rebound. Our contention is that each one of the elements ought to be studied with a different methodology. The prodrome may be examined with a psychodynamic approach since it deals mainly with the underlying fantasies behind and before the outbreak (Benaim et al., 1974). The epidemic moment is best served by an epidemiological approach since it concerns the clinical aspects of the distribution of the outbreak. The most precise and elegant study of this type was made by Moss & McEvedy (1966). Finally, the rebound can be studied with a social-historical approach because it deals with the contextual elements surrounding the outbreaks. The rebound is specially important to explain all that goes on in the maintaining of the episode from relapses to the social fate given to the outbreaks. De Certeau's (1970) studies of the possession of Loudun are focused in this third aspect.

We suggest that three aspects be taken into consideration in the study of the outbreaks: the context, the clinical aspects and the fantasy behind it.

When the context is overlooked or bypassed, reports only show an unintelligible collective emotional discharge (Smith & Eastham, 1973). Clinical factors also require careful description. It is important whether the outbreak shows up as a wave of anxiety, quasidissociative states or quasidelusional collective beliefs. The symptomatic picture is an interplay of twin aspects of a hidden fantasy: the cathartic expression of an affect linked with it or the mimetic expression of the fantasy. In collective anxiety states, the fear is preeminent and the fantasy remains obscured. We are close to panic states. In outbreaks presenting as dissociative states, both aspects are combined. The African episodes reported by Kagwa (1964) and Rankin and Philip (1963) on collective running show an underlying fantasy of being savagely attacked and contaminated by some atomic power. The response is a wild flight, a run against the time to come, that of cultural dislocation showing up at school where it is most acutely felt. The episode happened sometime after the signing of the Treaty of Moscow banning atmospheric nuclear experiments. This immediate context shadowed a much larger but diffuse threat and the symptomatic features of the outbreak illustrates both the fear and the underlying fantasy. Other episodes manifest themselves as monosymptomatic quasidelusional beliefs as reported by Medalia and Larsen (1958) and Johnson (1945). The mimetic expression of the fantasy is here preeminent at the borderlines of paranoid expression.

The last element of analysis is the underlying fantasy behind the episode. Frequent ones are ideas of pregnancy or sexual transgression as illustrated by Taylor and Hunter (1958), Knight et al. (1965), Benaim et al (1974), Teoh and Tan (1972). The collective aspect of the fantasy is based on identification with the wish of the index cases, and the traumatic aspect of the fantasy urging some repetition. This traumatic aspect often comes when some parts of reality in the triggering event feeds in the conflictual wish. Fantasies of passive and sudden aggression are also frequent: murders, poisoning, sudden death, etc.

When any of these three elements is absent from the analysis, something is missing in the understanding of the episode. Any one of these aspects is often over-developed according to circumstances or investigators.

This brings us back to summarize our discussion in Table 13.4 which follows.

TABLE 13.4
The Epidemic Process

Epidemic process	=	Prodroma	Epidemic Moment	Rebound
Descriptive elements	=	Fantasy	Clinical features	Context
Method of study	=	Psychodynamic	Epidemiological	Sociohistorical

The Metaphoric Model

The question of contagious psychopathology is still there unexplained. The metaphoric use of the word contagion leads us to an extension of the infectious medical model. One can ask: What gets transmitted? We would say a wish whose vector is a fantasy that circulates in a reservoir of susceptible persons. The susceptibility is defined by the conditions of participation to the group where the outbreak appears. Yet this is common to many groups and does not breed epidemic hysteria that often.

Epidemic hysteria is an acute group syndrome which serves to the abreaction of a conflicting situation as this is attested by many reports where the conflict stems from a fantasied (Champion et al., 1963; Jacobs, 1965; Johnson, 1945) or a real threat (Bell & Jones, 1958; Knight et al., 1965; McEvedy & Beard, 1970). Why does the conflictual situation provoke a maladaptive group response instead of such other reactions as the public analysis of the situation, the taking over by a leader or the development of organized confrontation? We would suggest three contributing elements: 1. the identification of the people in the group with the wish of index cases, 2. the regression fostered by the group situation, and 3. the unconscious or preconscious nature of the conflict or underlying fantasy. This last aspect is a crucial factor in fostering the manifestation of the conflict through displaced, symbolic or alternate channels. It therefore pulls upon the regressive situation as a weak point to manifest itself. The contagious process is then the shaping of a defensive social organization where the price to pay for being dysfunctional appears less than the upholding of the prior situation. This solution is still good enough: it assures a diffused responsibility between members, a displacement of wrongdoings and the taking over of the situation by external agencies (administrators, police, doctors, etc.). Moreover, it might be hypothetically put that the leader which operates as the Ego of the group aligns himself along the wish of the index case, thus fostering the identificatory process of all members to the wish of the affected individuals, increasing the fantasmatic resonance in the group.

The outbreak might also serve as a final common pathway in the interplay of various individual and group factors. This idea borrowed from neurophysiology suggests that identical symptomatic behaviors can express pathology caused at various levels by various etiological factors. This concept is useful insofar as groups have a limited range of ways to express themselves. Therefore the acute group syndrome can express both acute crises or chronic strain. Symptoms of the latter in industrial settings include absenteeism, fall in productive output, fast rotating staff, increase in labor union claims.

The Prodroma

One can derive more information about the subliminal beginning of epidemic hysteria by a closer study of the prodroma in suitable outbreaks (Sirois, 1977). Taking reports by Schuler and Parenton (1943), Taylor and Hunter (1958), Knight et al. (1965), Teoh and Tan (1972) and Benaim et al. (1974), one can show that before the epidemic moment a few people fall sick, usually three. When a first (A) person is "showing something," followed by a second and a third, then the outbreak shows up. The second person (B) often presents something peculiar: NZ speaks loud, SB cries, Betty calls on people, Louise and

Millie are isolated or unstable. It might be the one who experiences what the first person (A) represents at a fantasy level. The third person (C) is often strikingly normal and well accepted socially as if her role would be to mirror what comes from B as something acceptable. A preliminary hypothesis would be that this micro-group works like a dream process vis-a-vis some unconscious material. This formulation remains a tentative one. But the descriptive elements can be traced back in various reports. Moss & McEvedy (1966) in an epidemiological study says that before the outbreak something is going on: a first case, then another, later a few and the outbreak shows up. Old 19th century reports by Bouzol (1884) and Armaingaud (1879) suggest the same image.

The other structural elements of the epidemic process, the epidemic moment and the rebound also deserve a more carefully detailed analysis. This appears fairly complex, especially the passage from one element to the other. The rebound phase is best studied in large detailed monographic reports.

CONCLUSION

Some further aspects of epidemic hysteria were left out of this review. We will mention only the rarity of outbreaks reported in familial contexts (Calmeil, 1845; Dorfmuller, 1817; Neumann, 1899). Also the masking of outbreaks by social customs are illustrated by Douglass (1944). Finally, the transcultural aspects of the problem would deserve further study. The importance of motor symptoms in African outbreaks compares with the essentially different syndromes of latah or amok, where motor symptoms are predominant. We know that the so-called culture bound syndromes of Amok, Latah, and Koro also occur in epidemic forms.

In reviewing epidemic hysteria it was assumed that most of the reports were describing similar phenomena under various labels. The historical and geographical differences in the medical and cultural contexts were thought to account for much of the variance in the descriptions and nomenclature. The similarity of outbreaks involving school children or industrial workers appear strong enough to establish the core identity of the phenomenon. Other selected outbreaks in towns were probably metaphorically labelled epidemic hysteria. It reminds us again of the often diffuse clinical boundary of the syndrome.

The classical outbreak has the following epidemiological characteristics. It involves a small group of segregated young females. It appears, spreads, and subsides rapidly, with a bimodal curve of one group of early affected, most affected, often relapsing cases and frequently a second wave of least affected not relapsing cases. Symptoms are usually limited in time to one day. These are variable mostly fainting, malaise, convulsion with hyperventilation and excitement. Signs are scarce and laboratory values normal except for those associated with hyperventilating. Incubation is notoriously short. Adults are spared, with no secondary attack rate in household. Transmission is by sight or sound, fostered by reuniting the group.

It is brought about often by a trivial triggering factor which activates a larger but diffuse group tension relevant to its formation, task or maintenance. This group process carries important shared unconscious fantasies which are abreacted in the outbreak. When social controversial issues carry similar fantasies the outbreak can take endemic proportions.

REFERENCES

Armstrong, H., & Patterson, P. Seizures in Canadian Indian children. *Canadian Psychiatric Association Journal,* 1975, *20,* 247-55.

Armaingaud, Recherches cliniques sur l'hystérie; relation d'une petite épidémie d'hystérie observée à Bordeaux. Mémoire et Bulletin de la Société de Médecine et Chirurgie de Bordeaux, 1879, 551-79.

Bekhtereff, V. Données sur l'épidémie neuro-psychique observée chez les travailleurs d'usine de Riga et de Petro-grad en mars 1914. *Obozrienie Psikhiatrii Nevrologii.* (Review of Psychiatry and Neurology - Petrograd) 1914, *19,* 585-613.

Bell, A., & Jones, A. T. Fumigation with dichlorethyl ether and chlordane: hysterical sequelae. *Medical Journal of Australia,* 1958, *45,* 258-63.

Benaim, S., Horder, J., & Anderson, J. Hysterical epidemic in a classroom. *Psychological Medicine,* 1974, *3,* 266-73.

Bion, W. *Expériences in Groups.* Tavistock, London, 1959.

Bouzol, M. Relation d'une épidémie à phénomène hystérico-choréique observée à Algon (Ardèche) en 1882. *Lyon Médical,* 1884, *47,* 142-48; 174-84; 211-17.

Bozzolo, G. Tosse isterica epidemica. *Riforma Medica,* 1894, *4,* 735-37.

Bricheteau, F. Relation d'une épidémie de chorée à l'hôpital Necker. *Archives générales de médecine,* 1863, 6ème série, *1,* 433-47; 532-49. Paris.

Calmeil, L. F. *De la folie, considérée sous le point de vue pathologique, philosophique, historique et judiciaire.* Baillère, Paris, 1845.

Chambrelin, M. Note sur les contractures spasmodiques qui ont ·régne au pénitencier central des femmes à Namur depuis 1845. *Archives Belges de médecine militaire,* 1849, *4,* 169-200. Bruxelles.

Champion, F. P., Taylor, R., Joseph, P. R., & Hedden, J. C. Mass hysteria associated with insect bites. *Journal of South Carolina Medical Association,* 1963, *59,* 351-53.

Chew, P. K. How to handle hysterical factory workers. *Occupational Health Safety,* 1978, 50-53.

Coleman, J., Katz, E., & Menzie, H. The diffusion of an Innovation among Physicians. *Sociometry,* 1957, *20,* 253-70.

Colin, L. Une épidémie de possédés en Italie en 1878. *Annales d'Hygiène Publique et Médicale Légale,* 1880, 3ème sée. *4,* 5-10.

Cornish, J. On a convulsive epidemic in Cornwall, caused by a religious excitement. *Medical and Physical Journal* 1814, London. *31,* 373-79.

Cunliffe, W. J. *Psychic Possession among industrial workers. Lancet,* July 1, 1978, p. 44 (letter)

Davidson, A. Choreomania: An historical sketch, with some account of an epidemic observed in Madagascar. *Edingurgh Medical Journal,* 1867, *13,* 124-36.

De Certeau, M. La possession de Loudun, *Collection Archives.* Julliard, 1970.

Dorfmüller, H. Wahrnchmung einer Houslich Endemischen chorea St. Vitii und einer Chorea Epileptica von Insekten. *Zeitschift für Praktischen Heilkunde,* Nov. 1817, 101-108.

Douglass, J. H. The funeral of "Sister President." *Journal of Abnormal and Social Psychology,* 1944, *39,* 217-23.

Eastwell, H. D. A pica Epidemic. A price for Sedentarism among Australian Ex-Hunter-Gatherers, *Psychiatry,* 1979, *42,* 264-273.

Franchini, A. Manifestazioni isteriche collettive interpretate come sintomi di intossicazione da gas ignoto. *Medicina di Lavaro,* 1947, Italia, *38,* 57-60.

Friedman, T. I. Methodological considerations and research needs in the study of epidemic hysteria. *American Journal of Public Health,* 1967, *57,* 2009-11.

Gentleman's Magazine, 1787, *57* (lre partie).

Hecker, J. F. C. Epidemics of Middle Ages (Translated from German by B. Babington) London, 1844.

Ikeda, Y. An Epidemic of Emotional Disturbance Among Leprosarium Nurses in a Setting of Low Morale and Social Change. *Psychiatry,* 1966, *29,* 152-64.

Jacobs, N. The Phantom Slasher of Taipei: Mass Hysteria in a Non-Western Society. *Social Problems,* 1965, *12,* 318-28.

Johnson, D. M. The "phantom anesthetist" of Mattoon: A field study of mass hysteria. *Journal of Abnormal and Social Psychology,* 1945, *40,* 175-86.

Kagwa, B. H. The problem of mass hysteria in East Africa. *East African Medical Journal,* 1964, *41,* 560-66.

Kerckhoff, A. C., Back, K. W., & Miller, N. Sociometric patterns in Hysterical Contagion. *Sociometry,* 1965, *28,* 2-15.

Knight, J. A., Friedman, T. I., & Sulianti, J. Epidemic hysteria: a field study. *American Journal of Public Health,* 1965 *55,* 858-65

Levine, R. J. Epidemic Faintness and Syncope in a School Marching Band. *Journal of the American Medical Association,* 1977, *238,* 2373-76.

Lyons, H. A., & Potter, P. E. Communicated Hysteria—An Episode in a Secondary School. *Journal of the Irish Medical Association,* 1970, *63,* 377-379.

Maguire, A. Psychic possession among industrial workers. *Lancet,* Feb. 18, 1978. *18,* 376-78.

Martin, A. History of dancing mania: A contribution to the study of psychic mass infection. *American Journal of Clinical Medicine,* 1923, *30,* 265-71.

McEvedy, C. P., & Beard, A. W. Royal Free Epidemic of 1955: a reconsideration. *British Medical Journal,* 1970, *1,* 7-11.

McEvedy, C. P., Beard, A. W.: Concept of Bening Myalgic Encephalomyelitis. *British Medical Journal,* 1970, *1,* 11-15.

McEvedy, C. P., & Beard, A. W. A controlled follow-up of cases involved in an epidemic of bening myalgic encephalomyelitis. *British Journal of Psychiatry,* 1973, *122,* 141-50.

Medalia, N. Z., & Larsen, O. Diffusion and belief in a collective delusion: the Seattle windshield pitting epidemic. *American Sociological Review,* 1958, *23,* 180-86.

Michaud, L. Considérations psychopathologiques sur une épidémie d'hystérie convulsive dans un internat professionnel. *Archives francaises de Pédiatrie,* 1952, *9,* 987-990.

Moss, P. D., & McEvedy, C. P. An epidemic of overbreathing among school-girls. *British Medical Journal,* 1966, 2, 1295-1300.

Murphy, H. B. M. *Personal Communication* (January 16, 1974).

Neumann, M. Eine Hysterische Hausepidemia. *Monatschrift für Psychiatrie und Meurologie,* (Berlin), 1899, *5,* 405-410.

Ngui, P. W. The Koro Epidemic in Singapore. *Australian and New Zealand Journal of Psychiatry,* 1969, *3,* 263-266.

Nina-Rodrigues, P. Nouvelle contribution à l'étude des folies au Brésil. *Annales Médico-Psychologiques,* 1901, 8ème série. *13,* 19-31; 189-99; 370-81 et *14,* 5-18; 202-09.

Norman, E. H. Mass Hysteria in Japan. *Far Eastern Survey,* 1945, *14,* 65-70.

Parigi, S., & Giagiotti, F. Su di una epidemia di isterismo. *Rassegna di Studi Psichiatri.* (Siena), 1956, *45,* 1112-14.

Pfeiffer, P. H. Mass Hysteria Masquerading as Food Poisoning. *Journal of the Maine Medical Association,* 1964, *55,* 27.

Rankin, A. M., & Philip, P. J. An epidemic of Laughing in the Bukoba district of Tanganika. *Central African Journal of Medicine,* 1963, *9,* 167-70.

Rawnsley, K., & Loudon, J. B. Epidemiology of Mental Disorder in a closed Community. *British Journal of Psychiatry,* 1964, 110, 830-39.

Regnard, M., & Simon, J. Sur une épidémie de contractures des extrémités observée à Gentilly. Comptes Rendus des *Séances de la Société de Biologie* (Paris), 1876, 6ème série, *3,* 344-47; 350-53.

Robertson, F. Account of a singular convulsive affection which prevails in the state of Tennessee. *Philadelphia Medical and Physical Journal,* 1806, *2,* 86-95.

Rosen, G. Psychopathology in the social process. *Bulletin of History of Medicine,* 1962, *36,* 13-44.

Schatalow, N. Zur Frage von der epidemischen Hysterie. *Neurologische Centralblatt,* 1891, *10,* 405.

Schuler, E. A., & Parenton, V. J. A recent epidemic of hysteria in a Louisiana High School. *Journal of Social Psychology,* 1943, *17,* 221-35.

Seeligmüller, A. Ueber epidemisches Auftreten von hysterischen Züstanden. *Allgemeine Zeitschift für Psychiatrie.* (Berlin), 1876, *33,* 510-28.

Sirois, F. Epidemic Hysteria. *Acta Psychiatrica Scandinavica,* 1974, Suppl. 252 (Copenhagen).

Sirois, F. A propos de la fréquence des épidémies d'hystérie. *Union Médicale du Canada,* 1975, vol. *104,* 121-23.

Sirois, F. Remarques sur l'hystérie collective. *Evolution Psychiatrique*, 1977, *42*, 111-24.

Smith, H. C., & Eastham, E. J. Outbreak of abdominal pain. *Lancet,* Oct. 27, 1973, *2,* 956-58.

Stahl, S. M., & Lebedun, M. Mystery gas: An analysis of Mass Hysteria. *Journal of Health and Social Behavior,* 1974, *15,* 44-50.

Taylor, F. K., & Hunter, R. C. A. Observation of a hysterical epidemic in a hospital ward. *Psychiatry Quaterly,* 1958, *52,* 821-39.

Teoh, J. I., & Tan, E. S. *The Sociology of an Outbreak of Epidemic Hysteria.* Paper read at the 4th Conference on Culture and Mental Health in Asia and the Pacific, East-West Center, Honolulu, Hawaii, March 20-25th, 1972.

Teoh, J. K., & Yeoh, K. L. *Cultural conflict and transition; epidemic hysteria and social sanction.* Submitted for Publication to the Australian and New Zealand Journal of Psychiatry, October 1973.

Teoh, J. K., Soewondo, S., & Sidharta, M. Epidemic hysteria in Malaysian Schools: An illustrative Episode. *Psychiatry,* 1975, *38,* 258-68.

14 Stress and Coping Theory Applied to the Investigation of Mass Industrial Psychogenic Illness

Susan Folkman
Richard S. Lazarus
University of California, Berkeley

ABSTRACT

Coping is receiving growing attention as a critical process that mediates the relationship between stressful events and somatic, psychological, and social outcomes. This chapter suggests an approach to conceptualizing and measuring coping based on Lazarus' theory of stress and coping that can be applied to the investigation of mass psychogenic illness. In contrast to traditional approaches to understanding mass psychogenic illness in which chronic environmental conditions and presumably stable psychological tendencies are correlated, the approach offered here is transactional and process-oriented. Within this framework, determinants of coping are briefly discussed, and a metric for measuring coping is described. Findings from a recent study are used to suggest mechanisms through which coping might be related to mass psychogenic illness.

INTRODUCTION

One of the most interesting questions about mass industrial psychogenic illness is why some people become affected while others do not. Investigations of this question have been characterized by efforts to correlate illness on the one hand with *environmental factors* such as physical condition, organizational structure, and conditions promoting boredom, and *person factors* such as education, sex role identification, and pattern of defense on the other. This style of research measures presumably stable psychological tendencies and chronic environmental conditions, that is, structure rather than process. The goal is to make probabilistic statements of the sort: (1) certain types of people (with particular personalities and/or constitutional predispositions) are more likely than others to develop mass psychogenic illness; (2) any person is more likely to show mass psychogenic illness under certain kinds of environmental settings; (3) there is some form of interaction between (1) and (2) such that both sets of factors are necessary to produce mass psychogenic illness, or that both factors together increase the risk additively or in accord with some other mathematical function.

Although the structural research approach can define starting points by highlighting person-situation characteristics that seem related to vulnerability, it sheds no light on the psychological and social processes by which people become ill with psychogenic disorder. In this chapter we focus on one process in particular, *coping,* which findings from the laboratory (cf. reviews in Lazarus, Averill, & Opton, 1974; Lazarus & Launier, 1978), clinical settings (e.g., Cohen & Lazarus, 1973; review in Holroyd, 1979; Janis, 1958; Weisman & Worden, 1975) and work settings (e.g., Howard, Rechnitzer, & Cunningham, 1975) indicate is a critical mediator of the relationship between stressful events and somatic, psychological and social outcomes.

That there has been little coping research in psychogenic illness is not surprising, for despite the widespread importance attributed to coping in virtually every area of the behavioral and social sciences, little is known about coping processes, especially in adequately functioning populations. There are three major reasons for this state of the art. First, many investigations are motivated by an interest in the adaptive demands of highly unusual situations such as doctoral examinations (Mechanic, 1962), specific illnesses such as polio (Visotsky, Hamburg, Goss, & Lebovits, 1961) or cancer (Weisman & Worden, 1975), traumas such as burns (Andreasen, Noyes, & Hartford, 1972), or disasters such as war (Grinker & Spiegel, 1945). The purpose of these investigations is to understand coping that is adaptive and maladaptive in the specific situation of concern.

A second set of investigations is motivated by an interest in unusual groups of people such as those exhibiting some form of neurotic or psychopathologic behavior or who seem psychologically healthy, rather than ordinary people facing the normal trials and tribulations of living. Examples include studies of people who are phobic (e.g., Zane, 1978), angry (e.g., Novaco, in press), or schizophrenic (Zubin & Spring, 1977), and adolescents of outstanding competence (Silber, Hamburg, Coelho, Murphy, Rosenberg, & Pearlin, 1961) or "well-adjusted" college students (Grinker, 1962; Vaillant, 1977). In industry, too, research on coping generally focuses on a select group, namely, managers (e.g., Hall & Mansfield, 1971; Kobasa, 1979). Although there is research on stress and stress response in nonmanagement workers (e.g., Frankenhaeuser, 1975), there is an absence of information on the ways these people cope with their stressful environment. With the exception of the work of Pearlin and Schooler (1978) and our own current investigation, there has been little research about the coping processes of ordinary people who exist between the extremes of pathology on the one hand and "superperson" status on the other.

Third, research in coping has been severely limited by the measurement approach typically used. Most measures of coping are designed to assess a trait, that is, a stable personality disposition (e.g., Byrne, 1964; Goldstein, 1959). The purpose is to place the person somewhere on a bi-polar unidimensional continuum, and by virtue of this placement, to predict coping or other forms of behavior. Generally these continua reflect ego defense processes (cf. Haan, 1977; Menninger, 1963)—a somewhat narrow aspect of coping—anchored at one end by a "primitive" defense such as repression or denial and at the other by a "mature" defense such as isolation or intellectualization. Trait measures have not been very useful in predicting coping for several reasons. In personality research there is a growing consensus that knowledge of a trait is by itself not enough to predict behavior; there must also be information about the situation and its demands, and about the transactions between the person and the situation (cf. review in Pervin & Lewis, 1978). Second, naturalistic description of the ways people cope, gathered in clinical settings and from observations of the ways people handle real-life stressful events, shows that coping is not a simple, unidimensional process. It is a complex, shifting amalgam of thoughts and behaviors, some of which may be defensive in nature, but others of which can be restitutive or oriented towards problem-solving and mastery. This complexity is not captured in unidimensional trait measures.

What has been needed is a strategy of measurement that is appropriate in a wide variety of situations having different adaptational requirements, based on "normal" as well as disturbed people, and capable of describing the actual range of thoughts and behaviors people use in coping. We have been developing such an approach in our investigation of the ways community-residing middle-aged adults cope with the ordinary as well as unusual events of their daily lives. Over the course of a year, 100 middle-aged men and women were interviewed seven times about diverse, recently experienced stressful events. In each of the interviews subjects were asked to reconstruct what had happened, who had been involved, their thoughts and feelings during the encounter, its background, and how they coped with it. In the month intervals between interviews, subjects also filled out questionnaires in which they briefly reconstructed another recent event and indicated on a checklist the ways they had coped.

The coping checklist consisted of 68 items indicating various strategies used in the stressful encounter, including forms of direct action, information-seeking, cognitive coping activity (to avoid, minimize, or positively interpret an event), and inhibition of action (Folkman & Lazarus, 1980). At the end of the assessment period we had an average of 13.3 stressful events described for each subject, allowing us to examine the ways people actually coped with a variety of real-life situations, intraindividual patterns of coping across events, as well as normative comparisons based on situation and person factors. This process-oriented style of research, describing the ways people actually cope with events they appraise as stressful, is essential to help us understand the processes mediating between the person and the environment. Although an important goal in coping research is eventually to be able to talk about coping styles, it is first necessary to get a process-centered data base from which such abstractions can be made.

Over and above process-centered research strategy, a theoretical framework is also needed to organize findings and to guide further research efforts in a systematic, meaningful manner. The theory of psychological stress and coping developed by Lazarus 1966; (Lazarus, Averill, & Opton, 1974; Lazarus & Launier, 1978; Folkman, Schaefer, & Lazarus, 1979) provides a useful framework with which to consider coping and its relationship to mass psychogenic illness. Because it may not be familiar to many readers, we describe it in the next section before relating it to mass psychogenic illness.

A COGNITIVE-PHENOMENOLOGICAL THEORY OF STRESS AND COPING

We define psychological stress as a troubled relationship between the person and the environment in which an environmental demand, constraint or opportunity is judged to tax or exceed the person's resources (cf. Lazarus, 1966, 1980; Lazarus, Kanner & Folkman, 1980; Lazarus & Launier, 1978; Folkman, Schaefer, & Lazarus, 1979). This definition is offered within a transactional, process-oriented framework.

By transactional we mean that the person and the environment are always seen in *dynamic* relationship, each influencing and in turn being influenced by the other. The idea that stress refers to a relationship between the person and the environment has been expressed by numerous writers (e.g., French, Rogers, & Cobb, 1974; Klausner, 1971; McGrath, 1970; Mechanic, 1968; Van Harrison, 1978; Withey, 1962). However, the relationship is usually posited in static, structural terms. In contrast, we emphasize the dynamic, changing quality of the relationship.

When the person-environment relationship is conceptualized as dynamic rather than static, attention is shifted from independent person and situation factors to the *processes* through which the relationship is maintained or changed over time. Lazarus has identified two processes in particular, cognitive appraisal and coping, as critical mediators of this relationship. These processes shape the experience of psychological stress and the reactions to it.

Cognitive Appraisal

There are three types of cognitive appraisal, each of which serves in its own fashion to mediate the person-environment relationship: primary appraisal, which is the process of evaluating an encounter with respect to its significance for one's well-being; secondary appraisal, which is the process of evaluating coping resources and options; and reappraisal, which occurs as new information is obtained from internal changes, changes in the environment (often from coping efforts), from further reflection, and from cognitive coping processes.

Primary appraisal is directed at the question, "Am I okay or in trouble?" A situation can be evaluated as irrelevant or benign. Such appraisals are accompanied either by no emotion or by positive emotion. If, however, an encounter is appraised as involving harm (or loss), threat, or challenge, there is psychological stress. Harm/loss and threat appraisals are accompanied by negatively toned emotions such as fear, anger or guilt. A challenge appraisal marks an event as carrying the potential for mastery or gain, yet the person's adaptive resources are still taxed and as such it too involves stress. Challenge is characterized by the presence of predominantly positively toned emotions such as excitement or pride, even though negatively toned emotions might also occur (cf. Lazarus, Kanner, & Folkman, 1980).

Secondary appraisal evaluates the question, "What can I do about the stressful encounter?" and therefore it involves judgments about the accessibility and efficacy of coping resources. Such resources include general health and energy, good morale, social support systems and help-giving institutions, money, problem-solving skills, and meaning systems on which people can draw in sustaining their efforts.

Throughout the appraisal process, primary and secondary appraisal each influence the other. The extent to which a person feels threatened, for example, will be influenced not only by what is at stake (primary appraisal), but also by evaluation of available coping resources with respect to the requirements of the taxing situation (secondary appraisal). The more a person judges his or her coping resources to be adequate for a specific situation, the less threatening it will appear. The resources on which people draw—health, energy, morale, social supports and money—have been discussed elsewhere (for reviews see Berkman, 1977; Folkman et al., 1979) and need not be elaborated here. What we mean by problem-solving skills and meaning systems as resources, however, may not be as clear, and because they are relevant to our discussion of coping and psychogenic illness, a brief description of them seems in order.

Problem-solving skills include the ability to analyze situations for the purpose of identifying the problem, searching for information, generating alternative courses of action, weighing alternatives with respect to desired or undesired outcomes, and selecting and implementing an appropriate plan of action (Janis, 1974; Janis & Mann, 1977). Such generalized, abstract skills ultimately must be actualized through specific acts, in themselves often skills, such as changing a flat tire, managing a boat in a storm, presenting oneself to a prospective employer, preparing for an exam, and so on. Some writers conceptualize skills in global terms, such as dealing with moral dilemmas (Schwartz, 1970), with interpersonal problems (Shure & Spivack, 1978), with emergency situations (for reviews see Appley & Trumbull, 1967; Baker & Chapman, 1962; Coelho, Hamburg, & Adams, 1974; Janis, 1958; Lazarus, 1966), with role conflict (Katz & Kahn, 1966), with marital conflict (Levinger, 1966; Parsons & Bales, 1955) or with ambiguity (Haan, 1977). Others favor more concrete day-to-day actions such as one might use in a training manual (cf. Meichenbaum, 1977; Rogers, 1977; Yates, 1976). Both levels of abstraction are important for understanding and for practical intervention.

Meaning systems (e.g., beliefs about oneself and the world) are also highly relevant to coping. An important belief concerns a person's self-efficacy (Bandura, 1977). To disbelieve in one's own efficacy should generate passivity and disengagement, and at its extreme, should be associated with a sense of helplessness and hopelessness, which are in turn probably linked to depression (e.g., Engel, 1962; Lefcourt, 1976; Schmale & Iker, 1966; Seligman, 1975). This concept seems related to locus of control as discussed by Rotter (1966) and Lefcourt (1976), that is, the conviction that reinforcements are contingent on one's own behavior or, alternatively, are the result of external factors such as luck, chance or fate. The importance of self-efficacy as a coping resource has been demonstrated recently by Meichenbaum, Henshaw, & Himmel (1980) in a study of the cognitive strategies subjects used to tolerate the pain of holding their arm in a tub of cold circulating water (2°C). Tolerance of pain was not related to the cognitive strategies used to cope with the experience, but rather to the subject's general sense of efficacy in using the strategies required to cope with the stressor. Subjects who least tolerated the cold tended to see themselves as being overwhelmed and as having little ability to do anything to manage the pain.

Using a quite different approach, Stewart (1978) identified a set of beliefs having to do with the ways people organize experience, and demonstrated a relationship between these beliefs and the ways women cope with stressful events. She describes two contrasting patterns of cognitive organization: self-definition and social definition. A self-defining woman is relatively indifferent to sex norms (and perhaps to norms associated with other social roles), capable of emotional "distance" and objectivity, maintains a small number of intense, personal relationships, is active and interested in broad social movements and issues, and capable of vigorous instrumental activity. A socially defined woman behaves in accord with sex role norms, shows rather superficial emotional attachments to a large number of people, has little capacity to achieve detachment, is preoccupied with her own emotional and social life, and reveals a greater inclination toward expressive than instrumental activity. In a longitudinal study of 51 female college graduates, self-defining women were more likely than socially defined women to see their problems as external but the solutions as within their grasp (or internal), to see causes of problems as multiple, and to explain those causes clearly. In our terms, this type of self-related meaning or belief system should influence primary appraisal through the values and commitments expressed in the definition of sex role, and secondary appraisal through its impact on the development and use of instrumental, problem-solving skills.

These are just two of many meaning systems relevant to coping. Much has been written, for example, about existential meaning systems, such as faith in God, fate, or some higher natural order, that enables people to create meaning out of damaging experiences (cf. Benner, Roskies, & Lazarus, 1980; Bulman & Wortman, 1977; Dimsdale, 1974; Frankl, 1963; Moos & Tsu, 1977). Our purpose here, however, is to give examples of meaning systems that have particular relevance to coping with day-to-day demands.

A person's coping resources are usually not constant over time, that is, they are likely to expand and contract, some more erratically than others, as a function of experience and the requirements of adaptation associated with different events, styles of life, or periods in the life course. Furthermore, the presence in a person of a general resource such as problem-solving skills does not mean that the appropriate resource will be available or used at the time and in the transaction in which coping is required. Every major stressful transaction has its own task requirements: a problem on the production line makes different coping demands than problems with an aging parent. Even a person who is likely to be effective accross a variety of circumstances will be confronted by situations for which the requisite skills are simply not there or where some hidden personal agenda or psychological deficit will prevent effective use of available resources.

Some resources such as health, energy, morale and beliefs in one's efficacy seem to facilitate the use of other resources such as problem-solving skills or material goods. If, for example, one has learned problem-solving skills but does not have any conviction in his or her ability to use them, the skills will not be well utilized. Or if a person has a wealth of resources such as material goods, problem-solving skills and a general sense of self-efficacy but is low in physical energy and morale, the wealth of resources is less likely to be of much use. An interesting question concerns the mix of coping resources that one must have to meet the demands of various kinds of encounters.

Coping

Coping is determined by both primary and secondary appraisal. Primary appraisal is made in the light of the person's values, goals, commitments and beliefs, and generates information about the importance of the event to the person, its likely shape, and its potential implications for the person's well-being. This aspect of primary appraisal influences coping *directly* by affecting effort and mobilization. Generally, the more something matters, the more a person will put forth and sustain coping effort (cf. Atkinson, 1957; 1964; Klinger, 1977). Primary appraisal also influences coping *indirectly* through its impact on secondary appraisal; the definition of the task demands provided by primary appraisal guides the memory search, retrieval and evaluation of coping options that constitute secondary appraisal (cf. Folkman et al., 1979; Janis & Mann, 1977; Miller, Galanter, & Pribram, 1960). Secondary appraisal is thus shaped by primary appraisal and vice-versa. Convergence of the two appraisal processes determines the coping strategies a person uses and the effort he or she is willing to put forth.

Coping refers to "efforts, both action-oriented and intrapsychic, to manage (i.e., master, tolerate, reduce, minimize) environmental and internal demands, and conflicts among them, which tax or exceed a person's resources." (Lazarus & Launier, 1978, p. 311.) We are particularly interested in two types of coping because they have different but equally important functions. In the first, the person attempts to alter the troubled relationship with the environment by making changes in the person, in the environment, or both. We call this *problem-focused coping*. The second type is *emotion-focused coping,* which refers to efforts to regulate the emotional response to the problem. Emotion-focused coping includes cognitive and behavioral efforts to make oneself feel better through defensive activity, through expression of feeling, or through other "palliative" activities such as eating, drinking, smoking or talking with friends. These two coping functions have also been noted by others (e.g., Kahn, Wolfe, Quinn, & Snoek, 1964; Mechanic, 1962), and provide a useful way of describing broad coping functions in specific events.

Problem- and emotion-focused coping can be used simultaneously, as when a worker experiences anxiety while confronting a supervisor and needs to deal with both the anxiety and the precipitating problem. They can also be used sequentially as, for example, when a worker tries to get him- or herself "together," that is, emotionally under control, before an interaction with a supervisor that is anticipated as stressful. In both cases one coping function can facilitate the other. Throughout a stressful event, as emotions ebb and flow, a person may use now emotion-focused coping, now problem-focused coping, in some combination or other.

There are at least four basic modes of coping, namely, direct action, information-seeking, inhibition of behavior, and cognitive processes, that can each be directed at the problem or at emotional response (Lazarus & Launier, 1978). For example, direct action can be problem-focused, as in drinking, taking tranquilizers, or practicing relaxation or meditation. Information-seeking can be used to provide a basis for action directed at the problem, but it may also be used to bolster or confirm a decision that has already been made (cf. Janis & Mann, 1977), thereby regulating anxiety. Similarly, inhibition of action can be applied to both the problem and the emotion, as can cognitive modes of coping. In the case of the latter, cognitive efforts can be directed at problem-solving or at palliation. Further, each of the coping modes can be oriented towards events that have happened, that are happening, or that are anticipated.

Whether coping functions are classified into broad categories, for example, according to whether they are emotion- or problem-focused, or in the specific modes discussed above, the point to remember is that we are not talking about coping "styles" or "traits," but of the numerous processes of coping that a person engages in throughout a stressful event.

IMPLICATIONS FOR THE STUDY OF MASS PSYCHOGENIC ILLNESS

Psychogenic illness can be interpreted as an outcome of a chronic stressful situation that has been mediated by inadequate coping, the form of coping influencing the symptom pattern (cf. Menninger, 1963; Schafer, 1954; Shapiro, 1965). For coping to be successful, a person must:

1. control the situation instrumentally as effectively as possible so that he or she can move towards important goals or master obstacles; and/or
2. tolerate, minimize or accept damaging environmental pressures;
3. maintain value integrity and morale; and
4. keep the internal environment sufficiently close to homeostasis to maintain bodily health despite having to mobilize to cope instrumentally or with the emotional response (cf. Folkman et al., 1979; Roskies & Lazarus, 1980).

Often a person does well in one area, but pays a price in one or more of the others. For example, a worker might decide to increase his or her production pace to meet a quota, thereby being effective instrumentally, but do so at the expense of the family. This could cause anger or resentment towards the supervisor, guilt and/or loss with respect to the family, as well as fatigue. If similar transactions occur on a frequent basis, the continued mobilization necessary for both the instrumental response and for dealing with heightened emotions could lead to increases in hormonal secretions and eventually cause damage as described in Selye's (1946; 1976) General Adaptation Syndrome. Epidemiological research has also implicated mobilization in disorders such as hypertension (Ostfeld, 1967; Syme, 1977) and coronary heart disease (Rosenman, Jenkins, Brand, Friedman, Straus, & Wrum, 1975).

Relevant Patterns of Coping

Even though there are more females than males in the work forces in which mass psychogenic illness appears, still many more females than would be expected become ill. This suggests a gender difference in vulnerability and/or coping pattern. If coping is indeed related to somatic and psychological outcomes, as so much research indicates it is (see earlier comments), then a reasonable starting place for the investigation of coping processes in affected and nonaffected workers is to ask if there are gender differences in coping processes in the work setting.

There is a widely held belief that men are socialized to be instrumental, analytic and goal-directed in their thoughts and behaviors, whereas women are socialized to be expressive, compliant, dependent and deferential (cf. Broverman, Vogal, Broverman, Clarkson, & Rosenkrantz, 1972; Colligan & Murphy, 1979; Epstein, 1971; Ginzberg & Yohalem, 1966; Papanek, 1973). If these beliefs are transformed into coping terms, we would expect men to engage in more problem-focused coping than women and women to engage in more emotion-focused coping than men.

Our findings (Folkman & Lazarus, 1980) confirm these expectations. There were gender differences, but only in the use of problem-focused coping, and not in the use of emotion-focused coping. The gender difference in problem-focused coping, however, is limited to stressful events taking place in the work context; we found no gender differences in the use of problem-focused coping in the contexts of health and family. In other words, the obtained gender difference does not reflect a general disposition on the part of males to use more problem-focused coping than females in all settings, but only in the work setting. Here we have a good example of the importance of a transactional perspective of type of person (male-female) in commerce with a type of setting (work vs. health and family).

A similar pattern emerged when we examined gender differences in coping according to how situations were appraised. After reconstructing each coping episode, our subjects were asked to indicate which of the following statements best described the incident:

In general, is this situation one
1. that you could change or do something about?
2. that must be accepted or gotten used to?
3. that you needed to know more about before you could act?
4. in which you had to hold yourself back from doing what you wanted to do?

There were no gender differences in the use of these appraisals. Males, for example, did not appraise situations as ones that could be changed more often than females, nor did females appraise situations as ones that must be accepted or gotten used to more than males, as is usually assumed to be their sex role disposition. However, there were gender differences in the use of problem-focused coping as a function of such appraisals. Once again, men used more problem-focused coping than women, but only in situations that must be accepted, or in which more information was needed. And there were no gender differences in the use of emotion-focused coping as a function of the different kinds of appraisals.

These findings are particularly interesting in light of the kinds of problems that occur in work settings where mass psychogenic illness has occurred. These work settings are characterized by one or more of the following: highly authoritarian management styles, production pressures and unwanted overtime, physical stressors, boredom, and lack of two-way communication with peers and supervisors (cf. Colligan & Murphy, 1979). Such descriptions suggest that many workers are faced with work-related problems over which they perceive little personal control. Given our findings that men use more problem-focused coping than women at work and in situations where they perceive little personal control, and given that men are less affected by mass psychogenic illness, it is possible that the tendency to use problem-focused coping helps in some way with the psychological stress generated in these settings. Or, conversely, the disposition of women not to rely as much on problem-focused coping in the work setting increases their vulnerability or is reflected in somatization.

That men use more problem-focused coping than women at work, however, does not necessarily mean that the differences in coping are a function of gender. It may well be that women more often than men are placed in jobs where there are fewer opportunities to exercise personal control. If this is the case, differences in coping are not so much a function of gender as they are of the job. This is an important distinction, for measures to reduce susceptibility to mass psychogenic illness would differ in their emphasis if the determinant of coping is primarily the person (gender) or the environment (job). In the former case, for example, the emphasis could be on teaching skills for problem-focused coping. In the latter, the emphasis would be on changing the environment so as to provide more coping options for the worker.

Even though women are apparently more vulnerable to mass psychogenic illness than men, there *are* women who remain nonaffected. If problem-focused coping is related to psychogenic illness, then it is reasonable to ask whether nonaffected women use more problem-focused coping than affected women. Findings reviewed by Colligan and Murphy (1979) indeed suggest this may be the case. They cite Kerckhoff and Back's (1968) study in which affected workers were more reluctant to discuss work-related problems with their supervisors compared to nonaffected workers. Kerckhoff and Back also found that affected workers received little "information" support prior to an outbreak, although it is not clear whether non-affected workers received more.

With respect to emotion-focused coping, we found in our own research that more emotion-focused coping was used in situations appraised as having to be accepted than in situations about which it was felt that something could be done. Given the chronic nature of many of the problems of industrial workers, the role of emotion-focused coping as a mediator of psychogenic illness should be taken seriously and studied systematically.

A variety of specific coping activities are subsumed under emotion-focused coping. For example, a person can avoid thinking about a problem, minimize its importance, or engage physically in a diversionary activity (cf. Gal & Lazarus, 1975). One can also engage in emotion-focused coping by expressing feelings, using social suports to gain sympathy or understanding, interpreting an event in a positive light (Lazarus, 1981), or engaging in wishful thinking. In our study, people used an average of 14 emotion-focused strategies in each stressful episode they described. Although we have not yet analyzed the use of specific emotion-focused strategies or their patterns, the fact that they are so common as a class suggests that people actually use multiple modes in handling everyday stresses. The interesting question, then, is whether affected and nonaffected workers differ in the combinations or mix of coping activities and modes they use.

It is essential to look at the combinations of coping modes individuals use at different times during an encounter, in effect, as stressful encounters unfold. Earlier we identified four criteria that must be satisfied for coping to mediate a troubled transaction successfully. We also pointed out that coping that satisfies one criterion may be exacting a price in one or more of the others. In the same manner, a coping mode or pattern that satisfies one or more criteria at one stage of an encounter may not do so at another. For example, at the outset of a stressful encounter, denial can reduce psychological stress, but if its use prevents the person from taking action such as seeking needed medical care (cf. Katz, Weiner, Gallagher, & Hellman, 1970), it can prove costly in the long run. On the other hand, at an early stage in some stressful encounters (e.g., disabling illness), denial is often valuable by buying time until strength can be mobilized for rehabilitative efforts (cf. Hamburg & Adams, 1967). In the work setting, a person's feelings of frustration and anger can be reduced by minimizing the significance of troubling, boring or noxious aspects of a job that cannot be changed, thereby reducing psychological stress. In the long run, however, this can lead to alienation (Dubin, Champoux, & Porter, 1975; Hall & Schneider, 1972; Hall, Schneider, & Nygren, 1970; Kanungo, 1979) and be costly both instrumentally and to morale and value integrity.

LEVELS OF ANALYSIS

We have focused on two psychological processes, cognitive appraisal and coping, that we think play a role in symptom development and cessation in mass psychogenic illness. However, physiological and social processes are also involved in this phenomenon. Physiological processes are manifest in symptoms, and social processes are involved in providing some of the environmental sources of stress and in the "contagion" feature of mass psychogenic illness. The importance of all three levels of analysis demands that particular attention be given to how each level influences the other.

Research and theory in stress, coping and adaptational outcome (e.g., health/illness, morale, social functioning) occur at all these levels, but findings are confusing because the rules that link the levels are not clear. For example, the levels are often considered dependent or interchangeable, with the manifestation of stress at one level implying stress at another (Lazarus, 1966; Lazarus & Cohen 1976). Thus, it is assumed that if a person is showing physiological signs of stress, he or she is experiencing stress psychologically as well. However, a person may show marked physical stress reactions while jogging, but be in a happy, relaxed state of mind. Or, conversely, a person may show minimal levels of stress physiologically, but be observed as troubled psychologically, as in the case of a person who uses intellectualization (distancing, detachment) to deal with a difficult problem. In an even more complex example, and one that is particularly relevant to mass psychogenic illness, the onset of an illness (physiological stress) can bring psychological relief to a person who uses illness to escape a troublesome work situation while causing stress at the social (family or work) level. On the other hand, a decision to divorce may be threatening to a husband or wife at the individual (psychological) level and eventuate in illness, but relieve strain at the social (family) level by resulting in fewer interpersonal conflicts. Later, however, there may be cessation of stress at the psychological level when time has permitted effective adaptation, showing that all such relationships are time-contingent, varying greatly within it.

Although we must be wary of assuming that stress at one level implies an analogous process at another, often they occur together and, similarly, what is happening physiologically in the social stress context will not make sense without recourse to psychological mediators. It is important and legitimate to cross levels of analysis as long as these are dealt with and measured independently. Two recent studies illustrate how this can be done effectively.

The first example comes from some research of Pearlin (1975) who attempted to study a problem that has generated great interest in sociology, namely, that of status inequality between spouses as a source of stress in marriage. By means of interviews, Pearlin obtained information about the social status of both the husband and wife's parents (ascribed status) as well as about the importance of status to the respondents, some of the details of their married life and the amount of marital stress they experienced. In this way he was able to relate a social system variable, namely, inequality of ascribed status, to stress reactions via psychological mediation.

Pearlin's findings include three points of importance to us here. First, status inequality by itself did not produce stress in the marriage unless the higher-status partner considered status striving highly important (the psychological mediator), in which case that partner felt cheated. Second, status inequality resulted in subjective stress by disrupting four aspects of the couple's interaction, namely, reciprocity, exchange of expressiveness, exchange of affection, and the sharing of values. Third, in partners for whom status was important, status inequality created a lack of affection in the relationship, and inadequacy both in the communication and in the sharing of values. Pearlin (1975) concludes: "To account for marital problems and marital stress it needs to be recognized first of all that status inequality by itself is of little or no consequence. The importance of such inequality to marital problems and marital stress depends on the meaning and value that one attaches to it [p. 356]."

We can see in this research not only the mixing of levels of stress analysis but also how going back and forth between the social level and the psychological level clarifies what is happening. For example, the social variable, status inequality, does not by itself create stress; it only does so in those individuals for whom status striving (the psychological variable) is important. Without taking into account the psychological mediators accounting for individual differences (e.g., individual values, appraisals), we can see that the social variable would not be as useful as a predictor of stress in marriage.

The second example comes from research by Mason (1974) on endocrine profiles in monkeys and men under various physical stressor conditions including exercise, fasting, heat and cold. Mason sought to determine whether or not these divergent physical stressors produce a general endocrine reaction (in the fashion suggested by Selye's General Adaptation Syndrome), or produce reaction patterns specific to each type of stressor. From our present perspective, the most interesting feature of Mason's research was his attempt to "clean up the independent variable," as he puts it. That is, psychological threat reactions are usually confounded with physical stresses, so it is difficult to tell what it is that is producing elevated adrenal corticosteroids.

In exercise trials with young human males, in fasting trials with monkeys, and in heat and cold trials with monkeys, Mason took care to minimize novelty, uncertainty, extraneous psychosocial stimuli, and discomfort, in short, to reduce as much as possible any psychological threats that might be engendered by the physical procedures in the laboratory. For example, by minimizing competitive concerns and avoiding severe exertion, the danger that young men would be psychologically threatened in the treadmill exercise experiment was reduced; and by providing the subject monkeys with a nonnutritive placebo, they were less likely to be upset when bypassed by their trainer who provided food to the control monkeys in the fasting experiment; and in the heat and cold exposure experiments, temperature changes were made very gradually rather than suddenly so as not to alarm the monkey subjects.

Under these conditions of minimal psychological threat, it was found that each physical stressor produced its own unique profile of hormonal change (eight hormones were measured), and there was no general or nonspecific hormonal response to all stimuli despite expectations to the contrary based on GAS theory. Even more remarkable, when psychological threats were kept at a minimum, there was no elevation of the 17-hydroxicorticosteroid response except for the cold condition.

Here, as before, there has been a crossing over of two levels of stress stimulus analysis, the physiological and the psychological; only in this case, the mixing of levels had been hitherto inadvertent, the psychological level (threat) being confounded with the physiological (heat, fasting, cold, exercise). Such confounding appears to have misled previous researchers about the bodily response to various stress conditions.

The corticosteroid response appears to be particularly sensitive to psychological threat, which is often confounded with other, physical stressors (Lazarus, 1966; see also Lazarus, 1974). By systematically designing a study in which the contribution of each level can be separately distinguished, Mason seems to have substantially advanced our understanding of psychophysiological processes related to stress. The lesson to be learned is not only that intentionally mixing levels of stress analysis is desirable because it adds to our understanding, but also that the inadvertent or careless mixing of levels (where the two are not differentiated by stimulus manipulation and/or by measurement) can lead to very serious confusion about what is actually happening.

In the literature on mass psychogenic illness there is confusion due to a lack of awareness of the differences between structural and process-oriented research, and of the relationships among the physiological, psychological and social levels of analysis. For example, in a recent review of mass psychogenic illness by Colligan and Murphy (1979), reference is made to a number of multileveled factors such as boredom (a psychological *state,* not a condition of work), sex role identification (a psychological *trait*), physical stress (a physiological *state or process,* depending on how it is measured), interpersonal conflict (a social *process*), contagion (a social *process*) and a group of psychological and physiological reaction patterns including physical symptoms and subjective distress.

As we stated at the outset, structural and process-oriented research are both important in understanding mass psychogenic illness. However, care has to be taken to understand the nature of the variables being examined (structural or process), how to examine them, the information they can generate, and the place of information in the whole schema. For example, as we pointed out earlier, correlating gender and conditions of the work environment (e.g., organizational style) with illness outcome is a heuristic which can usefully narrow the focus of a study. However, these correlations do not tell us why some people become affected with mass psychogenic illness while others do not. To understand this, variables which shed light on psychological, social and physiological *processes* must be examined. Both forms of research are important; but the purpose of each style and the kinds of information they can generate must be clearly understood.

REFERENCES

Andreasen, N. J. C., Noyes, R. Jr., & Hartford, C. E. Factors influencing adjustment of burn patients during hospitalization. *Psychosomatic Medicine.* 1972, *34*, 517–525.

Appley, M. H., & Trumbull, R. On the concept of psychological stress: Some pervading issues. In M. H. Appley, & R. Trumbull (Eds.), *Psychological stress.* New York: Appleton-Century-Crofts, 1967.

Atkinson, J. W. Motivational determinants of risk taking behavior. *Psychological Review*, 1957, *64*, 359-372.

Atkinson, J. W. *An introduction to motivation*. Princeton, New Jersey: D. Van Nostrand, 1964.

Baker, G. W., & Chapman, D. W. (Eds.) *Man and society in disaster*. New York: Basic Books, 1962.

Bandura, A. *Social learning theory*. Englewood Cliffs, N.J.: Prentice-Hall, 1977.

Benner, P., Roskies, E. & Lazarus, R. S. Stress and coping under extreme conditions. In J. E. Dimsdale (Ed.), *The Holocaust: A multidisciplinary study*. Washington, D.C.: Hemisphere, 1980.

Berkman, L. *Social networks, host resistance, and mortality: A follow-up study of Alameda County residents*. Unpublished doctoral dissertation, University of California, Berkeley, 1977.

Broverman, I. K., Vogel, S. R., Broverman, D. M., Clarkson, F. E. & Rosenkrantz, P. S. Sex-role stereotype: A current appraisal. *Journal of Social Issues*, 1972, *28*, 69-78.

Bulman, R. J., & Wortman, C. B. Attributions of blame and coping in the "Real World": Severe accident victims react to their lot. *Journal of Personality and Social Psychology*, 1977, *35*, 351-363.

Byrne, D. Repression-sensitization as a dimension of personality. In B. A. Maher (Ed.), *Progress in experimental personality research*, Vol. 1. New York: Academic Press, 1964.

Coelho, G. V., Hamburg, D. A., & Adams, J. E. (Eds.) *Coping and adaptation*. New York: Basic Books, 1979.

Cohen, F., & Lazarus, R. S. Active coping processes, coping dispositions, and recovery from surgery. *Psychosomatic Medicine*, 1973, *35*, 375-389.

Colligan, M. J., & Murphy, L. R. Mass psychogenic illness in organizations: An overview. *Journal of Occupational Psychology*, 1979, *52*, 77-90.

Dimsdale, J. E. The coping behavior of Nazi concentration camp survivors. *American Journal of Psychiatry*, 1974, *131*, 792-797.

Dubin, R., Champoux, J. E., & Porter, L. W. Central life interests and organizational commitment of blue collar and clerical workers. *Administrative Science Quarterly*, 1975, *20*, 411-421.

Engel, G. L. *Psychological development in health and disease*. Philadelphia: W. B. Saunders, 1962.

Epstein, C. F. *Woman's place: Options and limitations in professions and careers*. Berkeley: University of California Press, 1971.

Folkman, S., & Lazarus, R. S. An analysis of coping in a middle-aged community sample. *Journal of Health and Social Behavior*, 1980, *21*, 219-239.

Folkman, S., Schaefer, C., & Lazarus, R. S. Cognitive processes as mediators of stress and coping. In V. Hamilton, & D. M. Warburton (Eds.), *Human stress and cognition: An information-processing approach*. London: Wiley, 1979.

Frankenhaeuser, M. Experimental approaches to the study of catecholamines and emotion. In L. Levi (Ed.), *Emotions—Their parameters and measurement*. New York: Raven Press, 1975.

Frankl, V. *Man's search for meaning*. New York: Washington Square Press, 1963.

French, J. R. P. Jr., Rogers, W., & Cobb, S. A model of person-environment fit. In G. V. Coelho, D. A. Hamburg, and J. E. Adams (Eds.), *Coping and adaptation*. New York: Basic Books, 1974.

Gal, R., & Lazarus, R. S. The role of activity in anticipating and confronting stressful situations. *Journal of Human Stress*, 1975, *1*, 4-20.

Ginzberg, E., & Yohalem, A. *Educated American women: Self-portraits*. New York: Columbia University Press, 1966.

Goldstein, M. J. The relationship between coping and avoiding behavior and response to fear-arousing propaganda. *Journal of Abnormal and Social Psychology*, 1959, *58*, 247-252.

Grinker, R. R. "Mentally healthy" young males (homoclites). *Archives of General Psychiatry*, 1962, *6*, 405-453.

Grinker, R. R., & Spiegel, J. P. *Men under stress*. New York: McGraw-Hill, 1945.

Haan, N. *Coping and defending*. New York: Academic Press, 1977.

Hall, D. T., & Mansfield, R. Organizational and individual response to external stress. *Administrative Science Quarterly,* 1971, *16,* 533-547.

Hall, D. T., & Schneider, B. Correlates of organizational identification as a function of career pattern and organizational type. *Administrative Science Quarterly,* 1972, *17,* 340-350.

Hall, D. T., Schneider, B. & Nygren, H. T. Personal factors in organizational identification. *Administrative Science Quarterly,* 1970, *15,* 176-190.

Hamburg, D. A., & Adams, J. E. A perspective on coping behavior: Seeking and utilizing information in major transitions. *Archives of General Psychiatry,* 1967, *17,* 277-284.

Holroyd, K. A. Stress, coping and the treatment of stress-related illness. In J. E. McNamara (Ed.), *Behavioral approaches in medicine: Applications and analysis.* New York: Plenum, 1979.

Howard, J. H., Rechnitzer, P. A., & Cunningham, D. A. Coping with job tension—Effective and ineffective methods. *Public Personnel Management,* 1975, *4,* 317-326.

Janis, I. *Psychological stress: Psychoanalytic and behavioral studies of surgical patients.* New York: Wiley, 1958.

Janis, I. Vigilance and decision-making in personal crises. In G. V. Coelho, D. A. Hamburg, & J. E. Adams (Eds.), *Coping and adaptation.* New York: Basic Books, 1974.

Janis, I., & Mann, L. *Decision making.* New York: The Free Press, 1977.

Kahn, R. L., Wolfe, D. M., Quinn, R. P., & Snoek, J. D. *Organizational stress: Studies in role conflict and ambiguity.* New York: Wiley, 1964.

Kanungo, R. N. The concepts of alienation and involvement revisited. *Psychological Bulletin,* 1979, *86,* 119-138.

Katz, D., & Kahn, R. L. *The social psychology of organizations.* New York: Wiley, 1966.

Katz, J. L., Weiner, H., Gallagher, T. G., & Hellman, L. Stress, distress, and ego defenses. *Archives of General Psychiatry,* 1970, *23,* 131-142.

Kerckhoff, A. C., & Back, K. W. *The June bug: A study of hysterical contagion.* New York: Appleton-Century-Crofts, 1968.

Klausner, S. Z. *On man in his environment.* San Francisco: Jossey-Bass, 1971.

Klinger, E. *Meaning and void.* Minneapolis: University of Minnesota Press, 1977.

Kobasa, S. C. Stressful life events, personality, and health: An inquiry into hardiness. *Journal of Personality and Social Psychology,* 1979, *37,* 1-11.

Lazarus, R. S. *Psychological stress and the coping process.* New York: McGraw-Hill, 1966.

Lazarus, R. S. Psychological stress and coping in adaptation and illness. *International Journal of Psychiatry in Medicine,* 1974, *5,* 321-333.

Lazarus, R. S. A strategy for research on psychological and social factors in hypertension. *Journal of Human Stress,* 1978, *4,* 35-40.

Lazarus, R. S. *The stress and coping paradigm.* In C. Eisdorfer, D. Cohen, A. Kleinman, & P. Maxim (Eds.), *Conceptual models for psychopathology,* New York: Spectrum, 1981.

Lazarus, R. S., & Cohen, J. B. *Theory and method in the study of stress and coping.* Paper presented at 5th W. H. O. Conference on Society, Stress and Disease: Aging and Old Age, Stockholm, Sweden, June 1976.

Lazarus, R. S., Averill, J. R., & Opton, E. M. Jr. The psychology of coping: Issues of research and assessment. In G. V. Coelho, D. A. Hamburg, & J. E. Adams (Eds.), *Coping and adaptation.* New York: Basic Books, 1974.

Lazarus, R. S., Kanner, A. D., & Folkman, S. Emotions: A cognitive-phenomenological analysis. In R. Plutchik, & H. Kellerman (Eds.), *Theories of emotion.* New York: Academic Press, 1980.

Lazarus, R. S., & Launier, R. Stress-related transactions between person and environment. In L. A. Pervin, & M. Lewis (Eds.), *Perspectives in interactional psychology.* New York: Plenum, 1978.

Lefcourt, H. M. *Locus of control: Current trends in theory and research.* Hillsdale, N.J.: Lawrence Erlbaum Associates, 1976.

Levinger, G. Sources of marital dissatisfaction among applicants for divorce. *American Journal of Orthopsychiatry,* 1966, *36,* 803-807.

Mason, J. W. Specificity in the organization of neuroendocrine response profiles. In P. Seeman, & G. M. Brown (Eds.), *Frontiers in neurology and neuroscience research. First International Symposium of the Neuroscience Institute, University of Toronto,* 1974.

McGrath, J. E. (Ed.) *Social and psychological factors in stress.* New York: Holt, Rinehart & Winston, Inc., 1970.

Mechanic, D. *Students under stress.* New York: The Free Press, 1962.

Mechanic, D. General medical practice in England and Wales. *New England Journal of Medicine,* 1968, *279,* 680–689.

Meichenbaum, D. *Cognitive behavior modification: An integrative approach.* New York: Plenum, 1977.

Meichenbaum, D., Henshaw, D., & Himmel, N. Coping with stress as a problem-solving process. In H. W. Krohne, & L. Laux (Eds.), *Achievement stress and anxiety.* Washington, D.C.: Hemisphere, 1980.

Menninger, K. *The vital balance: The life process in mental health and illness.* New York: Viking, 1963.

Miller, G. A., Galanter, E. H., & Pribram, K. *Plans and the structure of behavior.* New York: Holt, 1960.

Moos, R. H., Tsu, V. D. The crisis of physical illness: An overview. In R. H. Moos (Ed.), *Coping with physical illness.* New York: Plenum, 1977.

Novaco, R. A stress-inoculation approach to anger management in the training of law enforcement officers. *American Journal of Community Psychology,* in press.

Ostfeld, A. M. The interaction of biological and social variables in cardiovascular disease. *Milbank Memorial Fund Quarterly,* 1967, *45,* 13–18.

Papanek, H. Men, women and work: Reflections on the two person career. *Journal of Sociology,* 1973, *78,* 852–872.

Parsons, T., & Bales, R. F. *The family: Socialization and interaction process.* Glencoe, Illinois: The Free Press, 1955.

Pearlin, L. I. Status inequality and stress in marriage. *American Sociological Review,* 1975, *40,* 344–357.

Pearlin, L. I., & Schooler, C. The structure of coping. *Journal of Health and Social Behavior,* 1978, *19,* 2–21.

Pervin, L. A., & Lewis, M. (Eds.) *Perspectives in interactional psychology.* New York: Plenum, 1978.

Rogers, J. *You can stop: A smokEnder approach to quitting smoking and sticking to it.* New York: Simon & Schuster, 1977.

Rosenman, R. H., Jenkins, C. D., Brand, R. J., Friedman, M., Straus, R., & Wrum, M. Coronary heart disease in the Western Collaborative Group Study: Final follow-up experience of 8½ years. *Journal of the American Medical Association,* 1975, *233,* 872–877.

Roskies, E., & Lazarus, R. S. Coping theory and the teaching of coping skills. In P. Davidson (Ed.), *Behavioral medicine: Changing health life styles.* New York: Brunner/Mazel, 1980.

Rotter, J. B. Generalized expectancies for internal versus external control of reinforcement. *Psychological Monographs: General and Applied,* 1966, 80 (Whole No. 609).

Schafer, R. *Psychoanalytic interpretation in Rorschach testing.* New York: Grune & Stratton, 1954.

Schmale, A., & Iker, H. P. The affect of hopelessness and the development of cancer. *Psychosomatic Medicine,* 1966, *28,* 714–721.

Schwartz, S. Moral decision making and behavior. In J. Macauley & L. Berkowitz (Eds.), *Altruism and helping behavior.* New York: Academic Press, 1970.

Seligman, M. E. *Helplessness.* San Francisco: W. H. Freeman, 1975.

Selye, H. The general adaptation syndrome and diseases of adaptation. *Journal of Clinical Endocrinology,* 1946, *6,* 117–230.

Selye, H. *The stress of life (rev. ed.).* New York: McGraw-Hill, 1976.

Shapiro D. *Neurotic styles*. New York: Basic Books, 1965.

Shure, M., & Spivack, G. *Problem solving techniques in childrearing*. San Francisco: Jossey-Bass, 1978.

Silber, E., Coelho, G. V., Murphy, E. B., Hamburg, D. A., Pearlin, L. I., & Rosenberg, M. Competent adolescents coping with college decision. *Archives of General Psychiatry*, 1961, *5*, 517-527.

Silber, E., Hamburg, D. A., Coelho, G. V., Murphy, E. B., Rosenberg, M., & Pearlin, L. I. Adaptive behavior in competent adolescents. *Archives of General Psychiatry*, 1961, *5*, 354-365.

Stewart, A. J. A longitudinal study of coping styles in self-defining and socially defined women. *Journal of Consulting and Clinical Psychology*, 1978, *46*, 1079-1084.

Syme, S. L. *Psychosocial determinants of hypertension*. Paper presented at Hahnemann College's Fifth International Symposium on Hypertension, San Juan, Puerto Rico, January 1977.

Vaillant, G. *Adaptation to life*. Boston: Little, Brown, 1977.

Van Harrison, R. Person-environment fit and job stress. In C. L. Cooper & R. Payne (Eds.), *Stress at work*. New York: Wiley, 1978.

Visotsky, H. M., Hamburg, D. A., Goss, M. E., & Lebovitz, B. Z. Coping behavior under extreme stress. *Archives of General Psychiatry*, 1961, *5*, 423-448.

Weisman, A. D., & Worden, J. W. Psychosocial analysis of cancer deaths. *Omega: Journal of Death and Dying*, 1975, *6*, 61-75.

Withey, S. B. Reaction to uncertain threat. In G. W. Baker & D. W. Chapman (Eds.), *Man and society in disaster*. New York: Basic Books, 1962.

Yates, M. Coping: A survival manual for women alone. Englewood Cliffs, N.J.: Prentice-Hall, 1976.

Zane, M. D. Contextual analysis and treatment of phobic behavior as it changes. *American Journal of Psychotherapy*, 1978, *32*, 338-356.

Zubin, J., & Spring, B. Vulnerability—A new view of schizophrenia. *Journal of Abnormal Psychology*, 1977, *86*, 103-126.

Author Index

Italics denote pages with bibliographic information

Subject Index

D

E

T

U

V

W,X,Y,Z